The Building Site

The Building Site
Planning and Practice

JOHN M. ROBERTS

Landscape Architecture Department
Iowa State University
Ames, Iowa

A Wiley-Interscience Publication

John Wiley & Sons
New York • Chichester • Brisbane • Toronto • Singapore

UNIVERSITY OF
STRATHCLYDE LIBRARIES

Library of Congress Cataloging in Publication Data:

Roberts, John Mack, 1935-
   The building site.

   "A Wiley-Interscience publication."
   Includes index.
   1. Building sites.   2. Building—Contracts and
specifications.   I. Title.

TH375.R63   1983   690   82-21931
ISBN 0-471-08868-4

Printed in the United States of America

10 9 8 7 6 5 4 3 2 1

To Jack and Jo

Preface

It became obvious to me many years ago that the professions of architecture, landscape architecture, and engineering were not serving the public or private client as well as they could. These professions do not possess the ability *independently* to solve the problems or to meet the demands of a client coping with large, comprehensive site planning conditions.

As a practitioner, I had occasion to observe confusion, ignorance, and, in some instances, satisfaction as the various professions sought to apply their respective skills and attitudes to residential, campus, new town, recreation, housing, and commercial site developments—and I wrestled with the problems of contractors striving to achieve order from three sets of directives. Now, as a teacher, it troubles me to see student designers become devotees of their respective "truths" in single-purpose professions with little regard for or contact with interdisciplinary site planning.

Although this book was developed to answer questions about site development implementation, I trust that both students and practitioners will see the intertwining of the physical design professions. No one person should expect to have ready answers to all of the conditions set forth in these seven chapters. But, however incomplete the list of subjects or illustrations may be, there is a sufficient number to argue the case for a collective attempt on the part of these professions to meet the needs of their clients.

For each student, the text debates the constant pull between tradition and self-expression. Whenever possible, a figure and explanation will direct a traditional graphic or language expression for contract documents. More important, the directives are given reason through an explanation of why such a tradition exists—commonly as a means of maintaining consistency in communication or be-

cause legal tenets have forced certain understandings.

Do not presume that the text or I have any particular or extraordinary insight into law. No legal advice is given. Also, the general nature of a text such as this provides little opportunity to view legal principles with any specific application in mind. The best that I can do is to assist in relating contract documents to their fundamental reason for existence—to foster understanding among the parties to a design or construction agreement.

The text is intended to fit into the academic and site planning world in two ways. First, it can be used as a lone text for a professional procedures course involving general preparation of site contract documents and legal tenets. A second condition might involve this text as a supplement to other, more specialized texts offering in-depth treatment of construction law, site construction drawing preparation, specification writing, contract administration, site work, and the like, where an instructor wishes to explore both general and specific subject matter. For professionals, the text may assist the recent graduate in comprehending office practice and procedures, while the seasoned veteran may wish occasional support for a rusty memory.

My thanks to my wife for many hours devoted to typing and proofreading and for her never-ending support. Jeff Tyler, while a graduate student, assisted in preparing the majority of figures for Chapter 5. And special thanks are due for the cooperation of Jack Leaman, one of those thoroughly professional people who grace friendship.

JOHN M. ROBERTS

Ames, Iowa
March 1983

Contents

    7.10.4 *The Contractor,* **173**
    7.10.5 *Conditions of Guarantee,* **173**

7.11 Procedural Specifications and Guarantee, **175**
7.12 The Plant Schedule, **175**
7.13 Construction versus Horticultural Environment, **176**
7.14 Protection of Existing Plant Materials, **177**
7.15 Topsoil, **177**

    7.15.1 *Imported Topsoil,* **178**

    Selected Readings and References, **179**

    Index   **181**

The Building Site

Chapter One

Construction Documents and the Law

Without prejudice to creativity, there comes a time when one must stop suggesting and evaluating new solutions and get on with the job of analyzing and finally implementing one pretty good solution.

Robert Machol*

1.1 INTRODUCTION

Early in the twentieth century, it became painfully obvious that artisans knew very little about designing and constructing the physical infrastructure required by a nation bent on change. The artisan was then both designer and builder with very little need for advanced technology or precise communications. Agrarian life-styles had provided individuals with general experiences in landscape modification but such efforts did not necessarily produce advances in technology or procedures useful in the planning and construction of physical features. An urbanizing society and economic demands dictated that artisans learn how to build schools, canals, railroads, public buildings, parks, and other features of an urban infrastructure.

Forces that produced the physical growth of the United States simultaneously caused the decline of the artisan and introduced a division of labor within the construction industry. This division produced two allied groups. The first group, the designers, became chiefly concerned with the conception of ideals and, in the tradition of an earlier England, with acting as independent advisors on behalf of the project's owner. The second group, contractors and

trades, became concerned with the construction and implementation of the ideal. When functioning as contractors, this group may rely upon a description of the scope of work prepared by a designer. A contractor's work is generally documented in order to record the division of responsibilities among the various parties to a construction contract. In general, members of these two groups are specialists in their respective areas, each group having evolved procedures, technology, ethics, and law that can deal effectively with the increasingly complex problems related to today's site planning, architecture, and engineering.

Traditionally, the various documents that make up a construction agreement are composed of symbolic drawings and written words. Both designers and contractors have grown dependent upon these documents for communication. In many respects, written words and drawings are the only means of legal communication among the various contracting parties. Drawings portray the physical idea and ideal of the designer, provide a graphic view of the various parts and their assemblage, and define and record the scope of the work. Written words specify procedures and the performance of materials or equipment, define responsibilities, and narrate administration of the contemplated work. Together, drawings, written specifications, various contractual conditions, and a signed agreement form a set of contract documents that legally binds the various parties together.

*Machol, Robert "The Optimum Optimorum" *Interfaces*, 4, 4, Aug. 1974. With permission of the Institute of Management Sciences, Providence, Rhode Island.

The division of labor that created designer and contractor roles has also extracted a social and economic price to be measured by the success or failure of communications. Designers and contractors depend upon their ability to communicate. Designers, for instance, are dependent on the ability of contractors to translate, interpret, and then implement the ideas created by the designer. Unfortunately, a designer must deal in symbolic models, which contractors must use to form an image of what they must construct or install with sufficient accuracy to permit them to place a price on their work. Each essentially is dealing with graphic and written symbols that can only model reality. With such an interdependent need for understanding based on a primitive means of information exchange, ample opportunity exists for misunderstandings.

The possibility for misunderstanding is compounded by the way in which modern business is organized, with its intradivisions of labor. Contractors and trades divide their work into specialities, each with an assigned legal, guild, or union definition of its responsibilities. The sequence of communications is generally from prime contractors to subcontractors to the various trades. In other words, information pertaining to a specific item may progress through many people before it arrives at the person actually doing the work. Designers may also split their assignments into segmented responsibilities, with individuals assigned tasks requiring general to specific information. Some people design, others translate design to paper, others write technical specifications, still others administer on-site field operations. As with actual construction, ample opportunity exists within a design firm for communications to become misinterpreted or garbled among the various people involved.

Difficulty in communication underscores the variance between what a designer feels should be accomplished by a contractor and the frustrations experienced in attempting to communicate such ideals. The possibility of frustration and difficulty with dual communications is acknowledged by a paragraph that appears in similar form in almost every construction agreement.

> *The plans and specifications are intended to supplement each other so that any work shown on the plans and not mentioned in the written specifications, or vice versa, shall be as binding as, and is to be completed the same as, if mentioned or shown on both, and to the true intent and meaning of said plans and specifications. In case of conflict between the plans and specifications, the specifications shall govern.*

Such a paragraph reveals that a probability of error exists, that the location of any error is unknown, that exactly which of the two means of communication may conflict is unknown, that the true intent and meaning of the communication will be uncovered at some future time, and that drawings and written specifications function as one document. The paragraph is also symptomatic of the state of the art in communicating ideals to a contractor, inasmuch as the contractor is being asked to accept the responsibility for both locating and correcting errors.

1.2   SITE PLANNING

Site planning is the conscious spatial ordering of the physical features and the specific uses, technical implementation, and ecologic management of a particular portion of the landscape. A site plan records the scale and organization of physical features, locates specific uses, communicates the scope of construction and any modifications, and identifies specific zones of landscape management practices. Although the act of site planning may involve the necessary modification of a site to include architecture, engineering, and a restructuring of site characteristics, it does not depend upon the presence of either achitecture or engineering works.

A site may be defined by its specific use and anticipated modification within the foreseeable future. Detailed modifications can be identified and implemented as opposed to the ongoing formulation of suggestions as to what might be done. A sense of urgency and direct resolution of human uses prevails because implementation of the design concept is imminent.

A site and its environs become a given condition that controls or shapes the design efforts of the architecture, landscape architecture, and engineering professions. The magnitude and nature of the work are directly controlled by the characteristics of the topography, soil, weather, and vegetation, and by the shape and size of the site, zoning regulations, utilities, and off-site features. Each of the design professions is involved in site planning, design, and modification, either as incidental to their design profession or as a specialized concern.

1.3 PROFESSIONAL RESPONSIBILITIES

Whenever site design and modification require the esthetic and technical skills of design professionals, they become responsible to both the property owner and the contractor. The chief reason for this lies in a professional's acceptance of the role of agent. In traditional circumstances, a professional will act as an agent of the property owner. In an agent, there exist traditional ethical and legal concepts that place a designer squarely between the property owner and contractor, with the mandate of fairness to both parties.

It should be kept in mind that behind the discussions regarding designers and contractors lies an agreement for services that you, as a design professional, have signed, that is, you have given your word and placed your professional reputation on the line with a promise to a property owner that you will properly prepare and administer a construction contract in the name of, and as agent for, that owner. At best, if the question arises, a court will view your conduct in light of whether or not you saw the train coming and attempted to untie the owner's body from the tracks. In other words, did you use due care and skill in the execution of your professional responsibilities?

A designer, employed by a client as an independent contractor, ordinarily provides professional services as an agent. In a legal sense, a client and professional designer enter a fiduciary relationship, that is, one of mutual trust while engaged in a single-purpose endeavor. The client recognizes that a designer's credentials, experience, title, and license indicate competence to act as an agent. Mutual trust often is expressed as a meeting of the minds relative to esthetic and functional solutions to site planning problems. A client must perceive the agent as able and willing to act in good faith and in the client's best interest.

Several tenets of law, in general, can express the responsibilities and obligations required of a person acting as an agent. The first tenet concerns a client's assumption of responsibility for an agent's actions and words. In effect, an agent is retained to act in place of and to speak for the client.

Whenever a client–agent relationship exists, there will be a third party. In construction cir-

cumstances, the third party will be contractors, material suppliers, or others with whom the owner/agent will need to be involved to construct or install the work. In an agency relationship, a client gives expressed authority to the agent to act or perform certain duties. In addition, there exists a certain implied authority for the agent to act as necessary to protect the interests of the client. Third parties are thus allowed to seek redress of a grievance against an agent through action against a client who does not object to or prevent the agent's incompetent performance.

In most instances, relative to construction contracts, an agent receives expressed authority in the general or supplementary conditions of the contract documents. Through this mechanism, the client/owner gives the agent specific authority and obligations for the agent's actions during the construction and installation period with regard to third parties and the client. It is to be presumed that such authority and obligations are made acceptable by and through an agreement between the owner and client. Upon execution of a contract agreement between the owner and prime contractor, an owner defines the role of agent, the agent implies acceptance of expressed and implied obligations, and the contractor accepts the implied and expressed role of the agent to act in place of the owner.

A designer's role may extend beyond an agent's expressed obligations to third parties. Implicit in an agency relationship is a professional designer's duty to protect the client against conflicts with the health, safety, and welfare of the public, as implicit consumers, and the client, as an explicit consumer. In other words, clients and the public expect the design professions to be sensitive to their social, economic, and esthetic welfare as well as protective in matters affecting health and safety.

Another tenet of law that, in general, affects a fiduciary relationship includes issues of guarantee and warranty. An agent's role is one of trust and does not ordinarily carry with it a precise guarantee or warranty of perfection. Courts must often measure an agent's competency against some sort of standard behavior, usually defined as "due care and skill and reasonable effort." An agent does not, in effect, owe either the client or third parties a mistake-free effort, but does offer services based upon a reasonable use of specialized skills acquired by lengthy and comprehensive education.

A student of site planning will find that an agency role is not a one-way benefit to the professional's convenience and status. State licensing regulations, the courts, and public policy set implicit and explicit standards for professional conduct. Professionals in all fields accept a great deal of responsibility for their actions, care, skill, and reasonable attention to their client's needs. In return, the public gains a set of standards by which its courts and professions may measure and evaluate an individual's competency.

1.4   TRADITIONAL DESIGN AGREEMENT

The exact nature and scope of a designer's contractual responsibilities will be delineated by a formal or semiformal agreement between designer and owner, which will be private between the two parties. In general, the agreement might contain:

Date

Names of the contracting parties

A description of the professional services involved

Time limit, if any

How *remuneration* will be made, the fee

How the fee will be computed, the basis

When the fee will be paid

Who will pay the fee

Responsibility for design program preparation

A description of the *real property location and projected land use*

Information supplied by the owner and by the designer

Extra costs for travel, telephone calls, mileage, dining, etc.

Construction budget or cost estimates involved, and when and who prepares them

Nature of the designer's responsibility as agent

Extra work, services, and tests that may be required and their costs

Provision for *suspension*, abandonment, or cancellation of the agreement

Delineation of *ownership* of drawings, sketches, specifications, etc.

Costs of reproduction of drawings, specifications, etc.

Provision for *consent*, if an assignment of the agreement becomes necessary

Accounting records, if necessary to fee computation

Signatures of both parties, titles, authority to accept

Distribution of the accepted agreement

Each project will involve particular needs of the owner and designer as well as special circumstances necessitated by the site. In general, the agreement must delineate and define:

1. The intention of the *two parties to agree* and that they have come to a meeting of the minds and a mutual understanding.
2. That both parties are favored by the agreement and there is a *mutual advantage* expressed by a definition of the responsibilities of each party.
3. That the agreement has been *accepted* by both parties.
4. That *remuneration* in the form of a fee of some type has been given and that services to be rendered for that fee are stated.
5. That the contents of the agreement are *reasonable* to both parties within the time specified and with possibility of completion.

An agreement for services should be in written form. There are, however, several ways of doing this. If the proposed services and project are small enough to comprehend and develop in a one- to three-page business letter, this may be acceptable, and, for many clients, preferable to a legal-appearing contract form. However, there is no doubt that major services, lengthy time frames, and complex services should be covered by more complete conditions than may be possible within a business letter. For example, the American Institute of Architects (AIA) documents B141, B141A, B142, B142A, B151, or their equivalent, may be useful. However, many of the AIA standard forms of agreement do not seem to clearly express or define the types of services necessary to site work.

An agreement for professional services often is preceded by a proposal for those services. When a business letter is first signed and mailed by a designer to an owner, it is considered an *offer* to provide services. Only when an owner *accepts* the proposal (offer), by signature or otherwise, does it become an agreement. Published agreement forms, such as those of AIA, are also considered proposals until accepted by the owner. When a designer is on the receiving end of a proposal from an owner, for example, from a public agency or a company, it is the designer who must accept by signing the agreement. One tenet of law considers a person who writes an agreement to have a slight edge in understanding the contents of that agreement. Designers who compose their own agreements should consider that they will be at a disadvantage if the courts ever are asked to interpret such an agreement. In addition, it is almost always a court's contention that an owner is a layperson and interpretation is to be based upon the mutual intent of both parties at the time of an agreement's consummation, not during the course of the rendered services. This really means that if you do not make yourself clear at the beginning, there is very little that you will be able to do after an agreement has been accepted by both parties.

An agreement for professional services does not, in and of itself, establish any connection between the designer and a contractor. All an agreement can do is establish the fact that the designer will provide, for example, negotiation and administrative services for the owner and will act as the owner's agent. At this point, no contractor has been selected (at least in the traditional format). This subject is discussed specifically in Section 1.5.6, Contract Administration.

1.5 TRADITIONAL DESIGN PHASES

A site planner/designer has a professional vocabulary that differs in some respects from those of architecture and engineering. The following discussion centers around the typical terminology and the phases through which a site planning and design effort may pass. It is doubtful, however, that every project will proceed in the precise manner described, nor are the phases mutually exclusive and separate.

1.5.1 Concept and Schematic Phase

The conceptual design phase will consist of the gathering of owner-supplied program information and the assembling of base map and site information, during visits to the site and conferences with the owner. After a review of a design program and sufficient time, a designer will present several rough graphic line drawings that indicate the designer's initial site development concept. This phase will serve as a basis for discussions of alternative con-

cepts, budgets, timing, project scope, and the design program.

In general, this initial phase attempts to establish freedom for both the designer and owner in achieving a mutually agreeable design concept. Although the phase is extremely flexible and depends upon the nature of the project, it is critical at least to establish an understanding of just what services will be provided. If the project is quite small, this particular phase may take only an hour or so, while major projects can seem to last an eternity as the designer and owner strive to come to grips with how the site should look and function.

This is the idea phase. A designer might introduce a variety of alternative concepts verbally, as well as through "talking paper," in the form of sketches and quick concepts of possibilities. A designer may attempt too little at this point and be unable to portray an idea properly. On the other hand, finished drawings may end up in the wastepaper basket each time a new concept must be developed.

It is important at this stage, as well as later, that a designer not introduce, in writing or discussions, any guarantee or warranty of satisfaction. As previously mentioned, a designer's role does not necessarily include a guarantee or warranty of a client's satisfaction, economic gain, error-free service, etc. However, the designer can severely strain such a provision by making promises or taking actions that may explicitly or implicitly offer a guarantee or warranty of service.

1.5.2 General, Preliminary, or Master Plan Phase

The preliminary design phase constitutes a period of concept refinement documented with detail sufficient for structures, roads, walks, plant masses, and specialty elements to be located and scaled in relation to an accepted design concept and program. When necessary, a general concept of topography modification will be presented to indicate conceptual land forms. The designer will prepare, present, and discuss the construction budget, drawings, and sketches as necessary for the owner's understanding and acceptance of the design solution. It may be desirable to adjust the preliminary plan, construction budget, and design program in order to arrive at a mutually acceptable solution.

This particular phase of the site development is critical to both the client and designer. It is at this point that a design concept becomes fixed, both economically and within the minds of people. A mental image of the design concept is implanted and the final product must match this image or everyone will be disenchanted with the results. Needless to say, the fixing of a budget may make or break the final product. A designer must clearly delineate the final product and define a reasonable estimate of the costs.

The preliminary phase should be fairly flexible, in that a full discussion of the design concept is presented to the client and possible revisions debated. Of particular concern will be the organization of architectural features as they relate to pedestrian and vehicle circulation systems, topographic features, and vegetation. Once fixed, the preliminary plan will become the basis for preparation of contract documents and conceptual design changes will become very expensive to all parties.

In some instances, this particular phase may constitute the full services to be provided by a designer. When this occurs, the phase might be more appropriately referred to as a master plan or general plan phase. An owner may or may not carry out the concept or may improve only portions of the site over a period of time. Under such conditions, this phase may involve combinations of design programming, site selection, site analysis, ecologic studies, market studies, developmental phasing, environmental impact evaluation and studies, land use planning, and ecologic management.

1.5.3 Design Development or Precise Phase

Upon completion and approval of a site design concept and general plan, a more detailed phase of services will commence. Precise drawings will be prepared that will outline the scope and features of technical elements. These features may include, but are not limited to, grading and draining, plantings, utilities, irrigation, construction materials, construction details, outline specifications, and an updated construction budget. Issues concerning the quantity and quality of materials, their construction and installation procedures, relationships among the various design features, guarantees, bidding procedures, site management practices, and site maintenance will be discussed.

This phase of implementation might best be described as the nitty-gritty of the process. Up until this time, all of the discussions have been properly regarded as conceptual in nature—that is, flexible, debatable, subject to revisions and alternatives, as

idea stages, and evolutionary. However, the professional can play the game just so long before economics force firm decisions upon the designer and owner. Decisions that are made at this time become the basis for developing contract documents that soon follow.

No doubt, several conferences will be held between the designer and owner before all of the questions can be resolved. (Traditionally, this is true but some modern procedures of design may also involve a contractor or management people during this phase.) Basically, it is the responsibility of the owner's agent to explain alternatives, costs, and technical details. For example, the choice of paving materials must be related to architectural materials, costs, maintenance, and longevity. Plant materials should be discussed as to variety, form, color, costs, availability, planting seasons relative to the construction schedule, etc. The relationship between plant materials and an irrigation system's quality, capacity, scope, need, type, materials, operation, etc., must be resolved.

This phase is a prelude to the preparation of the complete contract documents. It should be clear to the designer and owner that failure to resolve issues can delay the design and implementation process, result in extra costs, or subject the owner and contractor to undesirable risks.

1.5.4 Contract Document Phase

Upon approval of the design development phase, contract documents will be prepared by the designer. These will consist of drawings, written technical specifications, and certain bidding documents sufficient and necessary to precisely set forth and document the entire scope of site construction and installation work. A final estimate of site development and construction costs will be prepared and approved. The owner will be responsible for the form of an agreement between the owner and contractor(s). The designer will assist the owner in the preparation of the general conditions to the contract and certain bidding documents. Legal assistance may be necessary in preparing many of these documents.

Usually, contract documents consist of drawings, bid documents, contract conditions, and technical specifications. Bid documents (various forms necessary to bid on the work formally) and the conditions of the contract (specific points that relate to the total management of the work and role responsibilities) are subject to preparation by different people, depending on the client, legal advice, and the complexity of the project. Contract documents are a means to an end, that is, they are the mechanism of communication that controls and measures responsibilities among people. Drawings depict the physical form, and diagram, plot, locate, relate, represent, and portray work to be accomplished. Words express work to be accomplished through terms or phrases with specific meaning in law, technology, or professional usage.

Technical specifications are written instructions to a contractor and are generally considered as containing information of the highest priority, often superseding drawings whenever a dispute over meaning occurs. However, technical specifications must be considered a part of the total contractual package. Technical specifications and other written portions of the documents are simply a second means of communication.

Bid documents are written instructions to a contractor explaining how a proposal is to be presented; presenting various requirements of the owner as to insurance, bonding, alternate bids, bid sums, terms of the agreement, payment schedules, and labor requirements; and providing sundry information on federal, state, and local requirements. Many of these documents will be "blank" forms to be executed after a contractor has been selected. Some of the forms are not a part of the agreement but are considered as information, whereas others are made a legal part of the agreement between the owner and contractor. In most cases, a measure of concern should be given as to whether or not legal counsel will be of value to the parties in preparing this information.

1.5.5   The Drawings

Location and Vicinity Plan

This plan identifies a site's location and property description in relationship to the local community. It is often coupled with a general legend and/or an index of drawings and used as a cover sheet for the drawings that follow in the set.

Construction and Site Plan

This is a scaled drawing of the entire site as one or several sheets. It locates, identifies, and indexes construction materials and details to specific portions of the site, and provides a contractor with an overview of the site, patterns, building locations, and other items that will modify the site (except plantings, utilities, and grading).

Existing Condition Plan

This plan identifies and describes the physical site as real property, and existing conditions such as topography, vegetation, buildings, paving, or utilities. The plan obligates an owner to provide such conditions as the time work commences.

Demolition and Site Preparation Plan

It defines, locates, and guides a contractor as to which portions of existing conditions must be protected, demolished, removed, or maintained prior to or during the course of the site work. It may involve the construction of temporary elements for use during the course of site work.

Dimensioning, Layout, and Horizontal Control Plan

This plan locates and identifies all proposed structural work, buildings, paving, walls, benches, fences, curbs, special features, and other elements necessary for precise positioning on the site. On occasion, elements of utilities, draining, lighting, and other important objects either may be located on this plan or positioned on their respective diagrammatic drawing. This plan is directly related to the real property description and existing site conditions.

Grading and Vertical Control Plan

It establishes finish elevations of all proposed construction and earth forms, delineates drainage directions by spot elevations or contour lines, relates buildings to their vertical position and elevation, and, occasionally, vertically locates positions of drainage structures and utilities. Erosion control measures (except plantings) would be located on this drawing.

Planting Plan

This plan locates and identifies plant species and their sizes, and may contain information as to existing plants and related management or maintenance work.

Irrigation Plan

The irrigation plan identifies the size and type of irrigation systems, including diagrammatic locations of pipe, sprinkler heads, and controls, and their respective performance characteristics.

Utility Plan

It diagrammatically locates and identifies existing and proposed utility systems throughout the site, such as electricity, water, gas, and telephone. It also identifies underground and above-ground installations.

Lighting Plan

This plan diagrammatically locates and identifies exterior luminaires with power sources, switching, and circuiting above and below grade.

Details of Construction and Installation

Each of the subjects mentioned ordinarily requires large-scale drawings that precisely delineate a specific portion of a structure or installation. Ordinarily called details, these drawings may take the form of plans, elevations, sections, or schematic diagrams that are larger than the site plan and of sufficient scale to specifically portray and describe materials, connections, fittings, joining, sizes, finishes, etc.

Schedules and Legends

The use of graphic symbols and abbreviated words is very much a part of contract documents. The technique of symbolizing meanings is well understood by the construction industry. Each of the symbols or abbreviated words will require a legend in order to completely define its meanings. Schedules are graphic charts that allow a short form of identifying and standardizing materials, sizes, manufacturers, etc., for specialty items.

1.5.6 Contract Administration

As previously stated, there is not, ordinarily, an agreement directly between a contractor and designer. However, in an agreement between an owner and designer there *may be* a responsibility to represent the owner during construction, that is, to check on the progress of the work, assist the owner in obtaining a contractor to do the work, guard the owner against defects in workmanship or materials on the job, certify payments from the owner to a contractor, and carry out other administrative assignments. For this activity to take place and for the designer to have any status on the site, the responsibilities of a designer are written into a contractor's agreement. Designer responsibility is usually spelled out in the general conditions of the contract. When a contractor signs an agreement, that contractor is simultaneously *accepting* the designer as the owner's agent with responsibilities as delineated within the conditions. In other words, responsibilities of a designer can be only those defined in an agreement between designer and owner *and* in the conditions attached to an agreement between an owner and contractor.

In general, a designer's administrative responsibilities and duties will consist of the following. However, an enterprising student and wary professional should receive a more detailed impression by reading a few "general conditions," such as AIA documents A201, A201CM, and A201/SC; the General Provisions of the General Services Administration, form 23-A; or standard state public works conditions. Each of these documents uses different terms to title the contract administrator and gives a designer varying degrees of responsibility in the following areas:

1. *To represent the owner* during the construction period and act on the owner's behalf.
2. *To visit the site periodically* and keep the owner apprised of the work's progress, guard the owner against defects and deficiencies in the materials or workmanship of the contractor, and ensure that the work conforms to the contract documents.
3. *To ascertain and approve payments* due the contractor for work completed.
4. *To interpret* the requirements of the contract documents and judge the performance of the contractor's work.
5. *To judge and render judgments* on certain matters in dispute between the owner and contrator.
6. *To review shop drawings* as necessary and guide their incorporation into the work.
7. *To exercise assigned authority* to reject the work or materials of the contractor when and if they do not conform to the contract docu-

ments, and to call for and interpret testing of the contractor's work within the bounds of the contract documents.

8. *To prepare change orders* and other documents that may be necessary to the orderly progress of the work and interpretation of the contract documents.

1.6 TRADITIONAL CONNECTIONS

Three persons or groups of people are normally involved in a traditional site construction effort. The *owner/client* is that person or group who pays the bills and needs assistance. In the case of site work, an owner owns or has rights in the real estate and is most often called the *owner*. A designer, as an individual or firm, usually, but not always, becomes involved when an owner needs advice and counsel. A *designer* is supposed to serve as the owner's agent, dream up clever ideas, and prepare whatever

documents may be necessary for an owner and contractor to reach an agreement. A designer is hired under an agreement that is distinctly separate from that with the contractor. A *contractor* is that person or group who constructs or installs the work, applies special knowledge to the project, and, surprisingly, is often able to figure out just what a designer had in mind. Sometimes the work proceeds smoothly, sometimes everyone understands exactly what must be done, sometimes the project is a success, and sometimes they all live happily ever after.

Figure 1.1 illustrates a traditional arrangement among the various parties. Note that those designers and contractors connected by an agreement directly with the owner are referred to as *prime designer* or *prime contractor*. Those persons connected by agreement directly to the prime contractor or prime designer are referred to as *subcontractors* or *subcontracting consultants*. Whether one is a prime or a sub depends upon the contractual connection with

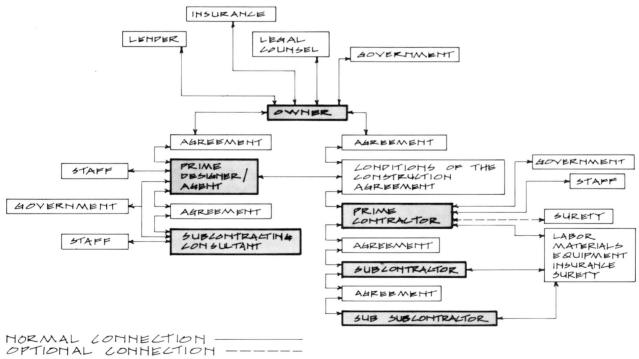

NORMAL CONNECTION ————————
OPTIONAL CONNECTION — — — — — —

MAY BE MORE THAN ONE PRIME
DESIGNER / AGENT AND PRIME
CONTRACTOR ON A PROJECT.

"TRADITIONAL" ARRANGEMENT PUBLIC OR PRIVATE

Figure 1.1 Typical and traditional relationship among owner, designer, and contractor.

the owner and not on what the job is about. Note that subcontractors are *not* contractually or otherwise directly linked with the owner.

Communication among the several contracting parties is generally and legally tied to their agreements. An owner and a prime designer communicate directly but a prime contractor must communicate with an owner through a designer, as ordinarily established by the general conditions of the construction agreement. Subcontractors communicate with their respective prime, and ordinarily never directly with an owner. Although these arrangements may appear cumbersome and dictatorial, they are necessary to precisely establish and maintain contractual responsibilities among the various people involved.

1.7   CHANGING TRADITIONS AND PROCEDURES

Reference to individual titles as defining an individual's responsibilities can be convenient and may simplify an explanation of today's construction industry, but such shortcuts tend to oversimplify reality. A much more complex condition exists than can be revealed by titles and flowcharts.

The past decade has seen the traditional roles of designer and contractor severely modified, and often obscured by contemporary circumstances. For example, changes in professional ethics, intertwining of corporate ownerships, revised modes of communication, and a steady increase in the complexity of contractual relationships have all contributed to a redefinition of traditional practice.

The sheer size and complexity of modern site planning projects have simply outgrown the traditional manner of development. The organization of traditional roles and practice firmly remains; however, traditional role definitions often fail to provide the ways and means to deal with many of today's site planning conditions. New procedures have evolved that combine rather than separate communications. Project management, turnkey, design/build, and fast track have all evolved to deal effectively with construction conditions requiring more efficient communications. Collaborative design and construction expertise, in many cases, have become necessary to design and construct large projects quickly. The evolution and return to almost an artisan format have been a direct testament to the need for construction and design speed. However, we find that the artisan of today often exists in the form of a corporation and its dual role of design and build becomes a managed team effort.

Professional ethics had for many years sought to separate a designer's role and income from an economic involvement with an owner. Design fees were based solely upon the provision of unbiased design services and set rather universally according to standards for services maintained by an entire profession. However, relatively recent consumers' attacks upon professional fee setting, and ethics, have forced design professionals to change their procedures with respect to fees and ethical responsibility to an owner. Designers may now choose entrepreneur roles involving both project design and possible economic gain from its success.

Many designers find themselves employed by both owners and contractors and suffering corresponding confusion as to their professional roles and responsibilities. Questions regarding exactly what constitutes correct practice and professional responsibilities continue to burden communications among all contracting parties.

At the present time, the only sure way of sorting out the responsibilities of various players is by identifying their respective contractual obligations and connections to each other. Titles may or may not truly explain a person's role or degree of, for example, responsibility as an agent to an owner. If a designer is an employee of an owner, the role of agent is very much submerged within the role of employee rather than being that of an independent professional. If a designer is employed by a contractor, the design professional's role with respect to acting on the owner's behalf may be nonexistent. Circumstances surrounding legal and ethical conduct must be weighed carefully by each designer because an individual's choice must often be made with limited or confused assistance from contemporary ethics.

1.7.1   The Contractor as Owner

Figure 1.2 illustrates a possible relationship between contracting parties whenever *an owner and contractor are one person or company.* Under this format, an agreement will exist between a designer and owner/contractor. Such an arrangement is possible for a single home owner or for thousands of acres under development by a large corporation. In general, the owner/contractor will let subcontracts for portions of the construction work and direct prime

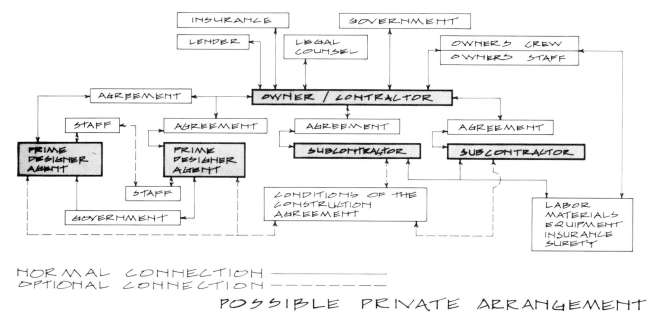

Figure 1.2 Typical relationship among principal parties to a construction contract whenever the role of owner and contractor becomes one personality, company, or corporation. Significant revision to communications occurs with designers connected to subcontractors by optimal owner–contractor contract conditions.

arrangements with a single designer or with several separate prime contracting designers. Note that the conditions to a construction agreement become an optional connection with respect to the various subcontractors and the designers. In other words, there may or may not be any traditional role of agent with respect to the designer(s) and subcontractors. This format requires careful attention to the itemization of the responsibilities of everyone concerned in any agreement for professional services.

Contemporary site planning projects may grow to such scale and complexity that owner/contractor roles are merged into one person or corporation. A designer's role and traditional location in the communication system may change drastically. The role may become specialized and may not involve the full complement of services traditionally reserved for the agent of an owner. If a site is large, several specialists in site work, architecture, and engineering may become independently attached to the owner through prime agreements. The owner/contractor and staff then become the filter through which all information travels. Each designer functions as an independent contractor for services. A designer's traditional role as observer of the work and manager of the construction agreement may be severely limited to that of a consultant with correspondingly limited vision of the entire project.

When an owner/contractor are as one, the prime designers must be aware, legally and professionally, of the exact scope of their duties and responsibilities as set forth in their separate agreements. It must be very clear as to the efforts and responsibilities of the owner/contractor's staff when coordinating among prime designers as well as the nature of communication among them. It is unfortunately quite possible for prime designers to function for years as single entities without any communication among themselves and with complete reliance on the owner/contractor for planning and programming information.

The traditional role of contract documents and designer responsibilities becomes obscure under a combined owner/contractor. Drawings and specifications are traditionally intended to describe work to be accomplished and also to protect an owner against fraud or misrepresentation by the contractor or the designer. In the case at hand, contract documents no longer must protect an owner unless the owner/contractor chooses to subcontract the work described by such contract documents. However, if the owner/contractor executes the work with staff forces, it is the owner/contractor who chooses the extent of the work costs, quality, and scope, and not the traditional agent. On the other hand, a professional designer cannot completely abdicate a design responsibility, because there must be concern for the public health, safety, and welfare. A contractor may not be accorded the same "layman" designation by law because of the contractor's more than casual understanding of the construction scene.

When the owner and contractor are one, verbal communication may take the place of the written or graphic instructions. Contract documents may be only superficially indicative of the work to be accomplished. Precise and detailed documents may be limited to those necessary to secure local governmental approvals or land use commitments rather than the traditional contract document role. In any event, the definition of services and responsibilities to an owner/contractor as well as the lines of communication will differ appreciably from the traditional concept of professional agency.

1.7.2 Design/Build

Figure 1.3 illustrates a typical design/build and turnkey arrangement. In this instance, a designer and prime contractor are connected as a company, by a special entrepreneur agreement in joint venture, or as a design consultant. The distinguishing characteristic is that a design/build group *works directly for an owner.* Under such an arrangement, an owner may, as an option, retain consulting designers, construction management, or testing firms to act as agents for the owner.

Basically, several historic and present-day circumstances have combined to give the design/build concept a strong competitive edge. With supreme court findings that the public consumer is best served by competition, ethics of the design professions have changed to encompass and allow a professional to function as a professional while both designing and contracting. The matter of ethics is philosophically cleared by informing an owner that the professional is, in fact, functioning both as a

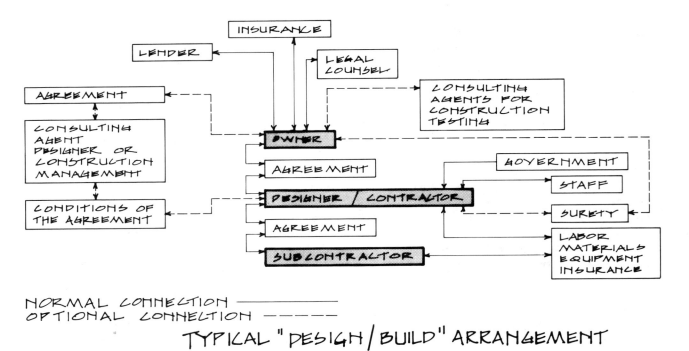

Figure 1.3 Typical "design/build" arrangement among parties to a construction contract. Note the absence of any traditional or "normal" connection between designer and owner. A designer's role as agent may become submerged within the designer–contractor relationship.

designer and as a contractor. To many designers, changes in ethical tenets have made design/build professionally attractive.

From the standpoint of an owner, design/build becomes attractive because the problems of coordinating either one or several prime design offices are avoided. The nature of design/build allows an owner to ignore the normal rash of decisions that usually occupy one's time during the design and construction phases.

The difference between consultant-prepared bid documents and design/build documents will be, basically, in the relative efficiency brought about by contract documents developed to fit the exact needs of a single contractor. Contract documents need not be issued to other contractors or, necessarily, to subcontractors. Their format may then be only what is seen as necessary to one contractor. For example, if a design/build contract calls for a complete irrigation system and the finished system and materials are specified and quality controlled by performance specifications, there need not, in fact, be any drawings prepared. Members of the design/build firm may simply design in the field and install the system using owner-approved materials. If the system fails, the contractor corrects the design until it meets the performance criteria.

When a practicing designer's client is also the prime contractor, lines of communication will be designated by the client/contractor. A designer may find that access to the contractor's client, the land owner, is limited or not allowed. A designer must then develop design concepts and communicate ideas only to the contractor who, in turn, may choose to or not to communicate those ideas to the land owner. Whether or not a designer performs traditional on-site observations of the work's progress and has any authority with respect to the quality of workmanship or materials must be explicitly identified within an agreement for services executed between the designer and the client/contractor.

1.7.3 Force Account

Traditionally the term *force account* is applied to the work forces of a governmental agency. The term can be applied to any work force employed by the owner or owner/contractor. A designer's ideas will be implemented by an agency's or owner's own work force. A designer's role may or may not be traditional in nature. For example, the construction documents may be developed only in terms of an agency's particular needs rather than being suitable for competitive bidding purposes. In addition, a designer may or may not be able to exercise any control of the design concept due to limitations of responsibilities in the owner–designer agreement. From a designer's standpoint, the use of a force account work force, either public or private, may lead to a loss of quality control and necessitate preparation of specially tailored construction documents unless the agreement for services allows otherwise.

1.7.4. Joint Ventures

Major site planning projects may necessitate the cooperative effort of several small but distinct design specialties. In most instances, the size of the offices, their individually limited staff or expertise, the scale of the project, and the respect among individual offices will allow them to function as a design team. Collectively, they can compete with large firms for design work that they could not handle as individual offices. Essentially, a joint venture is a temporary partnership that exists until the conclusion of a specific design project. As a type of partnership, each of the firms must clarify its responsibilities and services with regard to the project at hand. Individual offices are individually and severally bound to the owner/client for the performance of the agreed-upon design work.

A joint venture implies complexity and potential for confusion of communication. The several firms will be strained to set in motion quickly a communication system that may be quite abnormal to each firm's traditional way of operating. If, for example, the communication system does not have, or is unable to develop, a leader or filter for communication in an orderly fashion, chaos will be the net result. When offices are separated in time and space, the preparation of contract documents and their adequacy as communication with the contractor(s) will be severely strained. If the communication is broken at any point, the individual and collective firms may forfeit their reputations as well as economic viability.

1.7.5 Multiple Prime Designers

When an owner or contractor retains the services of several prime professionals, the designers make independent arrangements as to their scope of services, responsibilities, and fees. Unfortunately, each of the prime designers may then function independently and without coordination, except through the client. Such an arrangement is common with site planning, architecture, and engineering efforts for large site developments funded by private or public monies. Very often, a client will hire a management person to serve as an information and coordination source. The prime designers then become dependent upon the ability and desire of the client to provide and coordinate the project.

Personal experience in multiple design arrangements quickly reveals several problems that can arise among designers. Multiple and independent designers tend to see the site plan from differing perspectives and as functioning in various manners. For example, landscape architecture, architecture, and civil engineering all may become involved with a site plan during the conceptualization and implementation of their particular work. Each profession will view the site plan as it relates to that profession's scope of services and need for information. Each feels responsible for the site plan, and perhaps the owner fosters such an illusion by ignorance or a mistaken theory that a spirit of competition is good for a project's success. In any event, it serves no purpose to debate which of these professions needs a site plan because they all require such a plan for their work. What is debatable, however, is the usual lack of a single comprehensive site plan that contains all the information necessary to complete the project.

Unfortunately, the implementation of a project will usually proceed something like this: One of the design professions will become involved in the project first, develop an agreement, and possibly be followed by other prime designers as the client or designer faces specific problems. Each of the design professions will become responsible for a site plan in some form, but the first office will be called upon for information, drawings, concepts, etc., initially avoiding professional conflict. However, the first prime professional becomes increasingly aware of the fact that money and time are being expended in an effort to coordinate site planning and such activity was not contemplated as part of the original scope of services. In turn, the professions see their respective roles as exceeding their original scope of services or else find their design concepts thwarted by a lack of information or frustrated by a site plan they cannot control, or their concepts diffused within a team effort they did not anticipate.

With no single firm responsible, the owner or owner's staff perhaps ignorant of the problem, and several professions operating as independent specialists, no one ever puts the site plan into a comprehensive document. For example, an architect indicates concrete patio slabs extending 15 feet out from each building, a civil engineer delineates a drainage swale running parallel with, and 10 feet from, each building, a planting plan indicates trees planted on top of the engineer's electrical system, and a water system contractor just collapsed 200 feet of previously installed sewer pipe while dynamiting an excavation through rock. Such happenings occur in some form every day on every large site development.

There is no one answer to the situation. The examples only indicate a breakdown in the present system of proposals, agreements for services, and professional definitions of what constitutes professional practice. It is not so much that any of the professions have an obligation or a right to execute the site plan as it is the confusion of professional responsibilities that can result directly from multiple and independent design agreements.

1.7.6 Consultant, Independent Contractor, and Subcontractor

Designers should be aware of subtleties in language in their agreements that might place them in one of two liability categories—as an independent contractor (consultant) for consulting services, or as a subcontractor for design services. Although distinct liabilities are encompassed by each category, agreements may be unclear in distinguishing the difference.

A consultant or independent contractor is one who agrees to provide specific services to a prime designer and, in a legal sense, renders such service under the control of the prime in order to obtain a specific result while remaining free to use independent means to accomplish the results. The majority of agreements between a prime designer and consultant designer are intended to explicity prescribe a prime designer—independent designer relationship. Such a format allows both the prime contractor and consultant to be responsible for their professional performances. Although a prime designer must answer to an owner or third parties for the overall quality of total work, a consultant must also answer for errors or omissions that may occur while rendering specific services. In other words, an independent contractor for design services may be liable for acts, errors, or omissions resulting from a narrow scope of services.

Language or intent to develop a subcontracting relationship, while rare, explicitly recognizes a lack of privity between an owner and subcontracting designer. In essence, a prime designer retains full and broad control of the design work, either in fact or by actions, and may hold the subcontractor harmless or otherwise accept liability for the results of a subcontractor's narrow services, or somehow indemnify the subcontractor against actions by an owner or third parties. Under such a format, a third party's lawful action or recovery would be directed against the prime designer and indirectly against the consultant. However, a subcontracting designer may remain responsible to a prime designer with respect to ordinary application of due care and skill within the narrow scope of services rendered.

A central issue in defining or distinguishing between an independent contractor and a subcontractor is whether or not one party is able to control the means and methods by which a result is accomplished. Such questions often are asked in both designer and construction contractor relationships. In effect, a person (prime designer, prime contractor, owner) who controls both the means and methods and the results of a job will probably be regarded more as an employer than as a retainer of services.

1.8 SITE CONSTRUCTION AND INSTALLATION

Construction and installation for site work are contracted in much the same manner as architectural construction. However, the risk to all parties due to unknown site conditions presents unique problems. Many contractual formats are a result of engineering and landscape architectural conditions that recognize the unknown, or at least the degree of conjecture necessary to design and implement site work. For example, the quantities of earthwork, underground obstructions, weather, and whether plant materials will live or die become subjects of great risks to both owner and contractor. In many instances, it is either an educated guess, experience, or the artful application of the sciences that serves to estimate quantities and potential problems.

Designers must recognize the degree of risk taking for both owner and contractor(s). The gathering of insufficient site information will cost an owner excessive contingencies atop a fair price for the

contractor's work. On the other hand, a designer's fee will certainly escalate as more information must be gathered. Risk taking is a point of pride for many contractors but they also draw a line between potential speculative adventure and a recognizable disaster.

A designer should have some idea of the type of contractual arrangement that will be sought by an owner and the number of contractors to be involved, prior to executing an agreement for design services. Each of the various types of contractual arrangement will have some effect on the manner of contract document preparation. Construction arrangements will also affect the degree of construction administration and responsibilities required of a designer.

It is beyond the scope of this text to attempt a comprehensive discussion of the various procedures and legal ramifications involved with construction administration bidding, or the various documents that must be considered. The following brief discussion is limited to those points of procedure that directly affect site work.

1.8.1 Lump Sum or Stipulated Price

A lump sum arrangement provides an owner with one convenient construction price for the completion of all work directly and incidentally associated with the scope of work. Determining and programming the scope of work falls to an owner while the burden of describing that work falls upon a designer. A contractor is obligated to carry out and complete only that work described in the contract documents. Any changes in the scope of the work or the agreed-upon lump sum price can be accomplished only through a separate or amended agreement involving all three parties.

The nature of a lump sum agreement should limit its use to those projects that can be completely described in advance of any contractual agreement and completed by the contractor within a reasonable amount of time. Attempts to involve site elements with unusual risk factors will force a contractor to escalate the bid by a contingency factor equal to the risk being taken.

A lump sum contractual agreement is common practice because of its simplicity. However, it is often deceptively simple for many elements of site work. For example, when plant materials, soil quantities, or the nature of soils are difficult for a contractor to interpret, it may be best to separate these conditions from the lump sum price. Items separated from a lump sum price may be bid and priced as individual units to be computed and measured as may be required during the course of a contractor's work.

From the standpoint of legal framework, a lump sum price may prove to be a problem in a courtroom. Essentially, a contractor may contend that it was impossible to ascertain a reasonable estimate of lump sum costs because of hidden factors or factors that were impossible for a knowledgeable person to have foreseen. Even though most arrangements carry the provision that a contractor must visit, inspect, and determine those conditions that exist and bear upon the work price, it is not without precedent for courts to hold that the nature of a contractual arrangement was an unreasonable burden on the contractor and extensively in favor of the owner. On the other hand, it is reasonable to view a contractor as a knowledgeable person possessing sufficient skills to ascertain the scope of contractual work and accept contractual arrangements with eyes open. If a lump sum fixed price for work is subsequently revealed as being of undeterminable extent, a contractor may either execute the work and lose money or request recovery of unreasonable expenditures.

Many public works construction projects must be bid under a lump sum format or some other maximum cost arrangement. A fixed price is generally considered to be in the public's best interest. Some exceptions exist for the contracting of small projects, force account situations, or when emergency conditions exist.

A designer's duties in administering lump sum construction contracts encompass the reasonable application of due care and skill in preparing the contract documents and, in the main, in interpreting and clarifying the documents. The very nature of a lump sum arrangement presumes that all necessary information relating to the scope of work is contained in the contract documents.

A professional must be presumed knowledgeable in ascertaining the quantity and quality of information required by a contractor and to ensure an owner's anticipated scope of work. This does not necessarily mean that mistakes in documentation or misunderstandings as to the owner's wishes do not occur. It does mean that any conflict will be reviewed in light of a designer's actions and reasonable attempt to carefully and fairly document the value of a contractor's work.

Changes in the scope of a contractor's work are possible during the course of the site work but require a special written modification to an agreement called a *change order*. A change order is a formal amendment to an agreement between owner and contractor. Commonly, the amendment will contain a description of the addition, deletion, or modification of the materials or scope of the contractor's work and price, and any modification of the construction time. A change order must be approved by the owner, contractor, and designer. A change order is processed after an owner and contractor agreement is signed; thus its contents must be negotiated and are not subject to prior bidding.

1.8.2　Unit Price

This contractual method breaks down the scope of work into its component units. A contractor will submit a price for providing, constructing, and/or installing each separate unit as identified in a bid form. It is the designer who commonly prepares the form upon which bids are placed, with the approval of the owner. This type of arrangement is of particular value whenever the nature of site work is not easily defined or the quantities of work or materials are not determinable at the time of document preparation. However, sufficient information must exist regarding the characteristics of each unit so that a contractor may fairly determine its value and bid accordingly.

It is common engineering and public practice for a bid form to contain an "engineer's" estimate of the quantity of each unit; for example, linear feet (l.f.) of pipe, number of plants (ea.), the cubic yards of soil imported (yds.). There seem to be no universal standard units, and all parties must be aware of precise definitions of each unit as well as the possible use of metric units. A unit estimate is commonly and legally viewed as an estimate of quantity based on the best information available at the time of the agreement.

It is common practice to require the contractor to submit unit prices and then compute a lump sum or stipulated sum in the construction agreement. This procedure allows unit prices to be used in public works projects because it maintains a certain amount of nonnegotiated control over the addition or subtraction of units during the work's progress. However, the procedure has led to problems whenever a contractor's lump sum bid did not properly compute from the unit price data. Determination of a low bidder may be difficult when, legally, the lowest lump sum bid may not really be lowest when compared with an accurate computation of unit prices. A legal problem develops when determining which bid format takes precedence—a summary of the unit prices or the lowest lump sum.

Aspects of site work lend themselves very well to a unit price contract. Unit prices can allow the risk of unknown factors to be fairly spread between an owner and contractor. An owner will pay for the quantity of work necessary to accomplish a project while a contractor is given the opportunity to price the work competitively and receive reimbursement for each unit.

Caution must be exercised when deciding upon the quantity and characteristics of each unit. From the standpoint of a designer, aspects of project administration, unit measurement procedures, certification of payments to a contractor, and an owner's involvement must be assessed. Assuming that a designer's agreement includes administration of the work, the designer will be responsible for measuring each unit as complete and certifying payment for each unit within the context of periodic site visitations. If a unit must be measured on a continual basis, will the owner retain someone constantly to measure each unit? Measurement becomes a major problem to both designer and contractor. Both must protect their best interests.

Measurement of units should be planned at the same time a bid form is prepared. Payments to a contractor then can be based upon multiplication of individual units by a price. Wherever possible, the measurement technique should be explained in the contract documents so that a contractor is aware of and agrees to the procedure. Also, whenever possible, units that can be measured from drawings should be used as opposed to individual counts of units on site. Observable units are counted within other measurable units, for example, if plants are spaced on 6-inch centers, there will be four plants per square foot of area. Sprinkler systems can be measured as each sprinkler head, valve, etc., as a visible unit rather than as linear feet of buried pipe, that is, count rather than tape measure. In essence, units measured should be as large as possible with emphasis upon reducing time-consuming counting or one-site measurements.

1.8.3 Cash Allowance

The use of a cash allowance for minor work on construction projects has been a common practice for many years. Whenever an owner's indecision or a designer's nonresponsibility prevents an exact de-

scription of a portion of work, a stipulated amount of money can be estimated by the designer and included in a construction agreement. Work under an allowance usually is described in general terms so that the contractor understands its nature, magnitude, and relationship to other work. During the course of construction, an owner can decide upon the exact nature of the work and the contractor can proceed as directed. The precise value of the allowance is noted in the contract documents and becomes a part of the bid sum for the entire project. The owner's agent will provide normal periodic reviews of the work. For example, a stipulated sum of money can be noted on the bid form and added to a contractor's lump sum bid, for, say, five trees. When an owner locates the exact trees and so advises the contractor, the contractor will compute the real value of the work and negotiate with the owner regarding exact cost. The stipulated allowance sum is subtracted from the contractor's costs and the difference paid the contractor through a change order. Conversely, an owner may be credited money if the final costs are less than the allowance, again by change order.

Any cash allowance arrangement becomes critical whenever the scope of an allowance removes a designer from clear control of the work. There is, unfortunately, never a clear distinction between a minor and a major allowance figure. It might be argued, however, that such a distinction can be drawn whenever an owner's agent attempts to avoid responsibility through an allowance. Indecision on the part of an owner or nonresponsibility on the part of an owner's agent is uncontrollable, but a lack of professional knowledge, an uncaring attitude, and unethical conduct can be grounds for an owner to question a designer's judgment. Assume that an owner's agent has no knowledge of plant materials or their irrigation requirements yet agrees to include or guide the owner in such site work. The designer advises an owner that an allowance for the work can be placed in the contract documents. Upon approval by the owner, an allowance is written into the contract documents as a part of a contractor's site work. The allowance is termed "landscaping" and is defined only in general terms of the contractor supplying and installing plants with an irrigation system. As the scenario goes, the designer later approves the contractor's design, as well as general technical specifications for the work, and the contractor executes it. The question then becomes: Upon what body of knowledge did the designer base the approval of the contractor's design and how was the interest of the owner protected from misrepresentation of the work or of its value? Furthermore, is it illogical and unprofessional for an owner's agent to imply competency by guiding and approving such work when the agent obviously lacked sufficient skill and knowledge to clearly define and delineate such work in the original contract documents?

It may be helpful to judge the magnitude of a cash allowance independently of the total contract and related work. For instance, the example given might be related to the construction of a large building, and might at first appear to be a relatively minor percentage of the total construction contract. However, if the "landscaping" allowance were valued at several thousand dollars, the work could be independently viewed as major and critical to the owner's interest.

A contractor is usually allowed to design, to submit samples and prices, and to seek owner/agent approval for work under a cash allowance. Such work is not bid upon by a contractor. Any disagreement as to the value of the contractor's work becomes a matter for negotiation until a mutual agreement is achieved. Under such circumstances, the owner and designer usually are at a distinct disadvantage.

1.8.4 Incentives and Liquidated Damages

In principle, a contractual agreement exists for the length of time stated in the agreement or, if not stated, for whatever amount of time is reasonable. If an owner can gain from a speedy completion of the work, that owner may offer a reward as an incentive, or state damages that might ensue without a speedy completion. Damages that might accrue to an owner if the contractor does not finish on time are called liquidated damages. Such damages must be a reasonable estimate of value resulting from a delay in the completion of a construction contract. In essence, liquidated damages are agreed to as part of the construction agreement and represent a meeting of the minds among the parties.

Although these definitions may sound reasonable, they often inspire controversy, demand legal interpretation, and occasionally confound both designers and contractors. There is a fine line between what is considered a penalty and liquidated damages. Exact legal language and interpretation by a lawyer are often essential to make a distinction between the two. It is commonly considered that if a contractor is to be penalized for delay, there must also be an incentive or reward for early completion of the project. If liquidated damages are considered a fair estimate of damages, represent a true understanding of the parties, and relate specifically to a situation that will damage the owner through delay, they probably will not be considered a penalty. For example, an owner who cannot open for business on a publicly advertised date because of a contractor's delay may be damaged by such delay and recover liquidated damages as prescribed in the construction agreement. Usually such damages are deducted from construction payments due the contractor. On the other hand, the owner's circumstances and an agreement's wording may imply a penalty that will benefit only the owner. As with most contracts, courts tend to view an imbalance of benefits as not in the public interest, particularly when the owner is the writer of the agreement (through an agent) and the contractor must accept one provision among many. Under these conditions, it is in the best interest of both parties to balance a penalty with an incentive to reinforce the presumption that time is the essence of the contract and that construction delay will be injurious to the owner.

An incentive contract agreement is one that includes reward or incentive and is of benefit to both parties. This is fine in general, but can involve some difficult decisions for a designer involved in administering such a contract. Obviously the number of working days necessary to complete a site project are of interest to the contractor and the owner. Here the weather can make an incentive arrangement a nightmare for all parties. For example, it rains for two days. Are just two days lost? The answer is no, several more days will be lost because a landscape architect will not permit wheeled vehicles to work the wet soil, thus shutting down both building construction and planting operations. The actual number of days will depend on the amount of rainfall, the weather after the rain, the nature of the soil, and a defensible definition of when it is too wet for work. The landscape architect and planting contractor cannot allow wet soil to be compacted as it will affect the quality of their work and the contractor's guarantee. The building contractor can work only on inside jobs and no one is sure whether work by a partial crew should be counted as a full or partial day of work. Under these conditions, people tend to become very tense and they do not live happily ever after.

As an arbitrator of the agreement, the designer

must decide and define a contractor's compliance with regard to timely pursuit of the work. Exact information as to who caused any delay may be difficult to find, and even more difficult to prove in a court of law. Almost everyone will tend, at times, to delay a project's completion. Contractors usually have several jobs going at once and move people and materials around as necessary. Owners may delay in making decisions or paying their bills. Delays in a designer's approval of shop drawings, procedures, site visits, details, and installation of materials and equipment may also cause a job to move slowly.

Although it is an imperfect procedure, it does seem as though some type of incentive is necessary to assure timely completion of most site work projects. At least, specifying a range of time and setting forth the need for speed in the contract documents will alert the contractor, and provide valuable information to a court, if the problem goes that far.

1.8.5 Cost-Plus Arrangements

In principle, a cost-plus arrangement is a negotiated contractual agreement between an owner and a third party, the contractor(s). The agreement usually is negotiated either because an owner or agent is unable to fully document the scope of work or because the risk to a contractor is too great for the submittal of a lump sum price. In essence, the *cost* comprises the contractor's real cost to operate, and the *plus* is some sort of profit or other incentive beyond the contractor's cost.

When an owner's agent becomes involved in documenting or describing a cost-plus agreement, certain major factors will emerge as part of the conditions:

1. **Work.** A complete definition of the word *work* is necessary in order to determine the scope of an agreement. For example, the act of installing an irrigation system can be work, but are the contractor's supervision, procurement of materials, on-site design time, and negotiation also defined as work?
2. **Cost.** The word *cost* must be completely defined. For example, are such things as labor, materials, stationery, field office, telephones, travel, overhead, home office staff, and trucks to be considered costs?
3. **Profit.** How is profit separated from cost? What costs will profit be based on—that is,

how is profit computed? Profit might be a lump sum figure, a percentage of cost, or a standard multiplier factor of labor.
4. **Materials.** Who procures and delivers materials, how are discounts handled, who guarantees, who makes the selection?
5. **Equipment.** What about rentals, labor versus equipment, equipment on standby, rates of cost, servicing, repair, insurance?
6. **Revisions.** What is the effect on the agreement if changes in the project's scope are initiated?
7. **Weather.** What is the effect of contractor's standby in the event that weather conditions delay or damage work?
8. **Delays.** What about the timing of project completion, phasing, tentative schedules, weather, strikes?
9. **Designer.** What are the designer's role, obligations, and responsibilities for the project's administration and quality?

The use of a cost-plus construction agreement is a two-edged sword for a designer. On the one hand, some designers may view this as a golden opportunity to see their working concepts created almost overnight. Delays attributable to bidding and ordinary procedures for the documentation of work are reduced to a minimum. Decisions often are made in the field as the existing conditions and new work are analyzed and problems can be solved as they occur.

On the other hand, some designers may feel that they lose their identity as professionals, and so suffer from a status problem. In a large and fast-moving project, everyone tends to become part of a team effort as normal roles are adjusted to the task. Unfortunately, some designers discover that their mistakes catch up with them within a few days, instead of months, and the security of contract documents and a snug office is lost in the need to make immediate and correct decisions at the point of action.

Full development of contract documents may not be a normal procedure in a cost-plus arrangement. A lump sum contract may have been avoided because existing conditions could not be ascertained, the scope of the project was not clear, or the nature of the work was too risky for any contractor to attempt. Documents will be limited, therefore, to those necessary to get things rolling and for an owner and contractor to establish some sort of construction budget. There is a high probability that documents prepared by a designer will only establish an objective and convey ideas. For example, if plants are to be transplanted from their existing position to new locations, it may take only a conceptual drawing to establish a direction, with the final decision reserved for on-site assessment of plant availability, weather conditions, and the new locations for the plants. Of course, a designer or someone must maintain a complete record of a contractor's work and be able to certify the work quality and quantity.

1.8.6   Segmented Contracts or Multicontracts

Both the agent and contractor must be aware of an owner's intent to segregate portions of the work into multiple contracts. An owner's agent may be called upon to administer contractual terms for several contractors instead of perhaps one contractor. Conversely, the owner's acceptance of a low bid for each portion of the work is usually unfair to a contractor who bids on the work as a whole. The bid form should define exactly how an owner intends to handle the contract. Some published general conditions to an agreement allow an owner to accept either the whole or any portion of any bidder's offer. If such conditions exist and the bid form allows for sufficient breakdown of the bids, an owner may accept multiple contractors.

From the standpoint of working drawings, it is best that segregation of contractors be known in advance. Drawings may carry notations for several contractors regarding connections and fitting to each other's work but must not assume any responsibility for procedures.

1.8.7   Guaranteed Maximum Cost

Any cost-plus arrangements can be adapted to reflect a maximum or ceiling on construction and installation costs. However, the scope of the work must be readily identifiable prior to the contractual arrangements. The scope of work may be identified through the usual contract documents or described in written instructions. When maximum cost, sometimes called an *upset figure*, is coupled with an incentive arrangement, both the owner and contractor have an equal opportunity for success.

SELECTED READINGS AND REFERENCES

Abbett, Robert W. *Engineering Contracts and Specifications*, 4th ed. New York: Wiley, 1968.

Griffin, James M. *Landscape Management*. Redwood City, California: California Landscape Contractors Assoc., 1970.

Collier, Keith *Construction Contracts*. Reston, Virginia: Reston, 1979.

Hauf, Harold D. *Building Contracts for Design and Construction*, 3rd ed. New York: Wiley, 1968.

O'Brien, James L. *Construction Delay: Responsibilities, Risks, and Litigation*. Boston: Cahners Books, 1976.

Sweet, Justin *Legal Aspects of Architecture, Engineering, and the Construction Process*, 2nd ed. St. Paul, Minnesota: West, 1977.

Simon, Michael S. *Construction Contracts and Claims*. New York: McGraw-Hill, 1979.

Walker, Nathan, et al. *Legal Pitfalls in Architecture, Engineering and Building Construction*, 2nd ed. New York: McGraw-Hill, 1979.

Chapter Two
Working Documents

Entry-level positions with design firms are usually drafting oriented. By and large the drafting and production of quality working drawings is critical and respected but a lonely life requiring precise and detailed work conducted in the backwater of office life. It is possible to become a designer with all of the attendant fame associated with such talent, but only if you foul up often enough on working drawings.

John M. Roberts

2.1 DUE CARE AND SKILL

The economic and professional integrity of an office hangs in the balance each time an agreement for the preparation of contract documents is initiated. Economic integrity depends on the efficiency of an office staff to produce documents. Professional reputations depend on the documents' comprehensive, accurate, and clear delineation of a contractor's work. Attempts to speed the production of working drawings must be balanced against the need to retain graphic clarity and informational content. Procedures that generate contract documents that are unclear, incomplete, and inaccurate are not worth their cost.

In the final analysis, it matters little whether an office defines its drawings in terms of economics or graphics. What does matter is that drawings have been prepared with due care and diligence as defined by professional ethics and construction law. A drawing's worth will be measured in terms of its performance as a contract document, not as to its appearance or speed of production.

Design professions practice with due regard for the public's health, safety, and welfare. Public health and safety are a function of design. Public welfare is a function of a professional's ability to provide an owner and contractor with the mutual understanding and protection affordable through drawings and other contract documents. A professional does not guarantee the public or an individual owner error-free contract documents. However, a professional is expected by the public, individuals, and the courts to practice with the due care and diligence ordinarily expected of an expert in the design subject. Due care and diligence generally relate to an office's ability to prepare drawings and other documents that are accurate, concise, and clear.

2.2 COMMUNICATION

Contract documents are an abstraction of the real world. Drawings in particular attempt to convey messages to a contractor and an owner through symbols, notations, numerical measurements, legends, and abbreviations. As abstractions, graphics only symbolize reality. Understanding of graphic symbols is accomplished by the construction industry's ability to translate them. From a legal standpoint, contract laws often define the clarity of an agreement in terms of a meeting of the minds, that is, whether there was an understanding of obligations on both sides of an agreement. Drawings can only be an abstract model of what the parties to a contract expect to translate into reality. As such, they are a means to an end.

Tradition is a term used frequently with respect to the way in which things are done, or simply standard operating procedure. It is what people understand to be true. From the construction industry's point of view, tradition is a clear understanding of what the agreements and contract documents mean. Nontraditional aspects of the work might involve new abstractions, such as symbols, words, abbreviations, and delineations that are not clearly understood at first. Thus a designer must be extremely careful in introducing new abstractions that do not find common meaning in the industry. The novice designer, for example, often finds tradition boring, or stifling to the imagination. It may help, however, to understand that a tradition develops out of a need to maintain a common language. A designer cannot unilaterally change the meanings of symbols or words just to experience the joy of being different. If a designer's symbols are not understood, there is no communication. Without communication, there is no understanding. Without understanding, there is no meeting of the minds and no contract may exist.

However imperfect its means may be, communication is a basic necessity for a designer. The designer must communicate an ideal—that which exists as an imaginary solution to the problem at hand. This concept, an abstraction, is known only to the mind and is explainable only in the crudest of terms. If the ideal is not communicated, it may remain forever within the imagination.

2.3 ACCURACY

Accuracy applies to individual sheets as well as to an entire set of drawings. Because drawings are legally considered one document, items or directions appearing on one sheet are considered as appearing on all sheets and in the technical specifications. Information that is contradictory among sheets constitutes an error.

2.4 CONCISENESS

Although conciseness often is considered an economic gain for office efficiency, a concise set of drawings is also a boon to a contractor. Drawings that are concise seem a commendable goal, which is reached only to varying degrees by most offices. A major problem with conciseness seems to stem from a drafter's confusion as to just what is required and who will be reading the drawing.

2.5 OBSTACLES TO CONCISE DRAWING INFORMATION

2.5.1 Explanations

A common enough error stems from a drafter's mixing of design decisions with contractual obligations—for example, indicating a pipe diameter on an irrigation plan along with a volume of water flowing through a pipe. The anticipated volume of flow is necessary to the engineering aspect of pipe diameter but of absolutely no value to a contractor. Inclusion of water volume on drawings might imply that a contractor is responsible for volume. In this instance, the designer is responsible for the pipe size selection and is the only person concerned with volume. A contractor is responsible only for the pipe's installation. A design decision belongs in the office files and not on the drawings. Drawings contain only information that obligates a contractor and is action-oriented.

2.5.2 Future-Oriented Notations

Attempts to explain a future function or use of a particular subject are designer oriented. Such notes are useful and worth the time if they assist a contractor in identifying particular areas of the site. However, the use of such terms for categorizing, for example, overstory or understory trees, indicates the future growth habit of various tree species with respect to their eventual size. The terminology is of little concern to a contractor who is involved with the installation of plants that are essentially similar in size at the time of installation.

2.5.3 Design Intent Terminology

Intent terminology may be defined as words that attempt to describe what a designer would like to see but cannot define. In general, such words signal a designer's flexibility, and perhaps frustration, in communication. For example, such terms as good, clean, as directed, as approved, may, or desirable, acceptable, secure to, in the opinion of, rolling, incorporate, and either always seem to creep into contract documents. These words, when used alone, have one thing in common: *they are indefinable as contractual obligations.* Each word attempts to describe a designer's subjective image of what should be and will always lead to flexible subjective judgments. From a contractor's standpoint, such

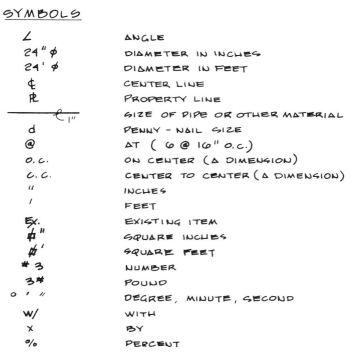

Figure 2.1 Typical language symbols traditionally useful as a shorthand means of communication. Many are so traditional that they are seldom explained by legend.

terminology conveys only the rawest idea of what might be expected because the real meanings are known only to a designer. Whether or not design intent terminology is fair depends on the nature of the contract agreement; that is, it may be unfair under lump sum but necessary under unit price agreements.

2.5.4 Abbreviations and Symbols

An abbreviation is a shortened version of a word, whereas a symbol might be defined as a graphic sign or emblem. In each instance, it is the meaning behind the abbreviation or symbol that translates into a contractual obligation, directive, material, or product. When abbreviations and symbols are defined within traditional usage, or legend, or schedule, the meaning becomes clear. However, it is unsafe to depend on common or traditional usage for a definition of meaning.

A novice designer sometimes hesitates to use abbreviations or graphic symbols to convey messages to a contractor. But the technique is traditional and regarded as proper, both economically and as a means of communication. There is no advantage and only time to lose if a designer avoids the extensive use of abbreviations and symbols. These shortcuts to communication free a drawing from excessive lettering and directives. A drawing is easier to read because it is uncluttered by extraneous letters and words. The drafter must decide whether it is faster to draw a symbol or abbreviate a word.

Symbols require careful planning. The key is to make them easily distinguishable from each other. For example, one must not depend upon size alone. The shapes of symbols should be designed to be distinctively different from each other so that they may be recognized visually. Before designing a new symbol, one should check to see that there is no traditionally accepted series of symbols for specific subjects. For instance, the plumbing, electrical, and welding trades each has its own symbols, which should be used to avoid confusion. A legend of symbols should be developed for use by the office staff during the preparation of contract documents, and a hierarchy of drafting difficulty established with relation to the quantity of symbols. The less graphically time-consuming symbols should be used most frequently on the drawings.

Symbols are used almost exclusively in electrical, lighting, irrigation, and utility system diagrams. Each symbol is identified and defined in a legend. With an irrigation system, for example, symbols are used to identify the location and types of sprinkler heads, the size and type of pipe, and the size and types of valves, meters, and vacuum breakers. The entire design is a diagram of the system as a whole with symbols identifying its various components.

2.5.5   Common Terminology and Specifications

Notes on drawings should be short and to the point with respect to the nature of materials, sizes, and particulars that also appear in written specifications. Written specifications are the proper place to precisely describe a particular material's characteristics, give measurements, and specify manufacturers or trade names. Drawing notations such as "Beam, Calif. Redwood, Const. Grade, S4S" should be reviewed in the context of what appears in other portions of the contract documents. If this is the only note and there are no redundant technical specifications, it is necessary. However, if the same information is contained in technical specifications or on a material schedule, the note is redundant, may contradict the technical specification, and consumes valuable drawing space. When specific information is contained in technical specifications or a material schedule, a drawing note should read merely "BM. RWD" (Beam Redwood), which assumes that only this particular beam is redwood. If all beams were redwood, there would be no need to repeat such information for each separate beam. If one carries this thinking a little further, one must wonder why the drawing is not sufficiently clear to visualize the fact that the wood member is a beam. Perhaps the note is not necessary as long as the size is indicated, say a 4 × 8, because a contractor will know what a beam is. If the drawing graphically describes the arrangement and size of all structural members, there is little need to note that this particular member is a beam. After all, a contractor is agreeing to install a 4 × 8 wood member as specified, not to install a beam.

2.5.6   Construction Notes

Most construction notes appear on drawings for two basic reasons. First, the drafter failed to convey such information to the specification writer and it is too late to place it in the specifications. Second, the information is extremely important to the success of constructing or installing that particular item appearing upon the drawings. The first reason is normal and a necessary use of the construction note. The second reason is often a figment of a drafter's imagination. Construction notes are generally an attempt to explain something to a contractor. They are seldom "extremely" important to anyone except the drafter. Construction notes developed for the second reason strongly indicate that a drafter is not familiar with a construction or installation technique and assumes that a contractor is also unfamiliar with it. Construction notes, in general, should be carefully worded so as not to void a contractor's responsibility for the means, methods, techniques, sequences, or procedures in connection with the described work.

When construction notes are used, they should not be scattered but grouped in one area of the sheet and identified as "Construction Notes."

2.5.7   Overdrawing

Overdrawing refers to extraneous lines, words, and textures that are incidental to contractual information or legibility of the drawings, and may develop from a drafter's attempts to be artistic with a purely functional contract document. Texturing, for example, tends to veil a drawing. A drawing often appears to be an academic exercise in drafting technique rather than to be delivering a message in a simple and concise manner. Extraneous or superfluous line work includes such items as grass or other vegetation with realistic detail, graphic soil texturing that reduces the clarity of dimensions, and planting plans with tree branch structures indicated.

2.5.8   Legends and Schedules

Legends and schedules have much in common, yet are separate in their use in documents. A legend is generally viewed as a means of matching a graphic symbol with its meaning. For example, an irrigation legend identifies each symbol used on the plan with its respective manufacturer, size, and performance. A contractor reads each symbol's meaning from the legend. In some instances, it may be appropriate to include the number and location of a detail that indicates how each item is to be constructed or installed.

Although not a requirement, it is common courtesy to place the legend or schedule on the sheet

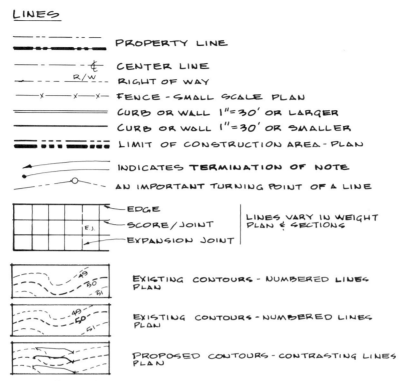

Figure 2.2   Typical symbolic language and meanings defined by lines. Lines vary as solid, broken, thin, and wide, and may be combined with language symbols in Figure 2.1. Note that meaning is implied both by the character of a single line as well as by the relationships among several combinations of lines.

that carries the subject plan view. A legend or schedule may also be bound in the project manual, but this practice is generally inconvenient for the contractor in the field (however, it does force the contractor to have both drawings and specifications on hand).

A schedule differs from a legend only in that a schedule does not usually contain graphic symbols. Its most common uses are in wood construction and planting. For example, a schedule of a wooden deck might prescribe that joists, beams, and posts be of a certain size and material. A plant schedule may be accompanied by general symbols that represent plants and their location on the plan. Under such conditions, it might be proper to call it a legend, but there is little to be gained in such a debate.

When legends and schedules are used on drawings, there is an opportunity for redundancy to occur. It is a mistake, for example, to include the same information both in a legend and in the technical specifications. One of the documents should predominate. A schedule is a graphic means of indicating typical elements, that is, all elements of the same size. An error arises when there may be a few atypical items on the drawings that are not noted.

Schedules and legends are a sure means of achieving conciseness in drawings. Rather than having notations of information scattered throughout, each type of schedule centralizes and reduces notations. Schedules also lend themselves to graphic portrayal of items, such as soil boring sections, soil profiles, plant forms or branching patterns, or special treatments for specific items.

2.5.9   Meaning of Lines

Presentation drawings, at the concept phase, may be embellished as artwork. Contract drawings are limited to lines that represent a physical entity. For example, only one line contractually describes the edge of a physical item whereas a presentation drawing may not even show a line or definite edge. Every line appearing on a contract document must have a meaning in reality or be otherwise contractually definable in fact.

Only two basic lines exist for a drafter's use—solid and broken. Solid lines can vary in width and density with pencil but may vary only in width when ink is used. Broken lines vary considerably as to break locations, width, and density. There are traditional messages in the type of line and how it is used. Unfortunately, the novice designer too often opts for experimentation with lines. For example, the use of a solid line in place of the traditional alternating long and short sequence of a centerline can be very confusing to the reader. In other words, the reader expects a particular line to have a certain message. In the case of nontraditional use of a solid line as a centerline, the message might imply the edge of paving rather than a centerline. The meaning of a line depends upon what the surrounding lines appear to be. If, for example, the line indicating the edge of paving and the centerline are the same width, have the same value, and are solid, who can blame anyone but the drafter for misunderstanding the message? Failure to communicate can be attributable to ignorance of line symbols, to pure laziness because it's easier to draw a solid line than a broken line, or to experimentation by a drafter who thought it would be fun.

2.5.10　Texture

Graphic texture is used on both the plan and details of an object or material, but only to distinguish symbolically among various adjacent materials. Texturing must not be considered an excuse for artwork.

One must distinguish among textural symbols used in plan, in elevation, and in sectional views. For example, the graphic texture of concrete in section might symbolize aggregate whereas in plan or elevation views the texture might symbolize surface. Particular attention must be given to details where inconsistent or incorrect use of a material symbol could translate into construction using the wrong material. If graphic texturing is to be used, the drafting team should decide exactly what symbols will be used. Textural symbols should be drawn, a standardized legend prepared, and copies distributed to everyone at work on the drawings. Again, there are traditional symbols that should be used, but the symbols for texture often must be invented.

Graphic texturing is a common, and often inefficient, product of the pretentious and lazy drafter. Overtexturing is a part of overdrawing. Quite often, the overzealous use of texturing masks important information such as dimensions, notations, and directives.

2.6　CLARITY

Many of the same items mentioned as being a deterrent to concise information are also a factor in drawing clarity. In many instances, clarity is obtained simply by the drafter's delineation of information in an accurate and concise manner. Clarity refers to the legibility of information on each drawing and among drawings. A drawing's clarity will also depend to a large extent on the drafter's skill in line work, lettering, and organization of the information. A factor often ignored at the drafting board stage is that many site drawings are used out-of-doors. What appears to be clear on a drawing board is not the same as when one is trying to hold the sheet on a windy day and to read the small print through specks of dirt on the hood of a pickup truck. Lettering must be large enough to be read without missing letters and, if necessary, to be legible after photo reduction. Line work should vary, particularly in width, as related to the importance of the delineated site object.

A most flagrant disregard for clarity is found in the amount of information included on each sheet. It is commendable from the standpoint of reducing reproduction costs to save a sheet or two of drawings, but the result often confuses the drafter as well as the contractors. It may not always be cost efficient, but each sheet should contain only one information subject rather than being cluttered with information regarding many subjects in an attempt to use all available space.

2.7　PROCEDURES

2.7.1　Sepia Reproducibles

Multiple subjects common to site planning lend themselves readily to the time-saving and repetitional aspects of sepias. Basically a single base or plot plan is prepared on a transparent paper or film medium and a series of sepias reproduced by a light-sensitive process or photography. Each sepia then becomes a base upon which each subject drawing may be drafted. This technique results in a saving in drafting time as well as the avoidance of discrepancies between the repetitions' base information. Sepia reproducibles will accept either pencil or ink drafting. The material may be a transparent polyester film or paper and "erasable" or "stan-

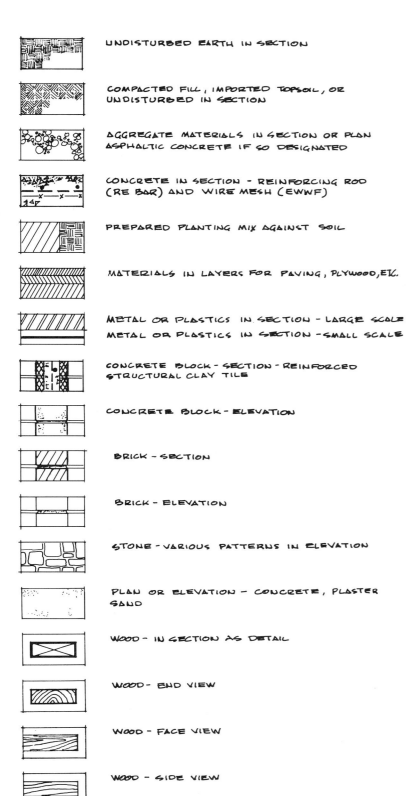

UNDISTURBED EARTH IN SECTION

COMPACTED FILL, IMPORTED TOPSOIL, OR UNDISTURBED IN SECTION

AGGREGATE MATERIALS IN SECTION OR PLAN ASPHALTIC CONCRETE IF SO DESIGNATED

CONCRETE IN SECTION - REINFORCING ROD (RE BAR) AND WIRE MESH (EWWF)

PREPARED PLANTING MIX AGAINST SOIL

MATERIALS IN LAYERS FOR PAVING, PLYWOOD, ETC.

METAL OR PLASTICS IN SECTION - LARGE SCALE
METAL OR PLASTICS IN SECTION - SMALL SCALE

CONCRETE BLOCK - SECTION - REINFORCED STRUCTURAL CLAY TILE

CONCRETE BLOCK - ELEVATION

BRICK - SECTION

BRICK - ELEVATION

STONE - VARIOUS PATTERNS IN ELEVATION

PLAN OR ELEVATION - CONCRETE, PLASTER SAND

WOOD - IN SECTION AS DETAIL

WOOD - END VIEW

WOOD - FACE VIEW

WOOD - SIDE VIEW

PLYWOOD - FACE

Figure 2.3 Examples of typical but regionally variable textures used to distinguish various materials from each other. Note that some textures are used as plan and elevation views while others delineate only materials viewed in "cut" sections.

dard." To erase the sepia, a special bleaching liquid is usually necessary, but *some* success can be achieved with moistened rubber or electric erasures, depending upon the material.

Figure 2.4 traces several plan sequences, which must be planned prior to beginning working drawings. Steps 1, 2, 3, and 4 relate to (1) preparing base map information, (2) reproducing several "sepia reproducibles," (3) drafting specific information on each copy of a sepia and (4) reproducing prints. Steps 1, 2, 5, 6, 7, and 9 relate to (1) and (2), preparing a base map and several sepia reproducibles, and (5) each sepia can be returned to the drawing board for the addition of information necessary to a particular subject, (6) a second generation of sepias can be reproduced from the reworked sepias in a quantity matching the numbers of finished working drawing sheets, (7) each second-generation sepia is returned to the drawing board for the addition of specific information and finish, and (9) prints are reproduced. Steps 1, 2, 8, 9, 10, and 4 relate to pin bar and photographic reproduction methods. One or several base maps are prepared at step 1. Step 2 becomes the photographic combination of the several base maps and production of sheets suitable for the addition of specific data at step 8. A second photographic combination of various sheets may be necessary at step 9 and drafting is finished at step 10. Prints are reproduced by photo- or light-sensitive

methods at step 4. The "original" base map(s) may be retired to file, and later use, at step 11.

The base or plot plan of a site must be carefully analyzed as to exactly what physical elements should and should not be initially drawn on the "original." Elements usually consist of existing and proposed architecture and/or structural slabs, walls, property lines, and the like, which must be reproduced on each of the subject sepias. Title blocks may be partially completed with those items of information that must be included on each of the subsequent sepias. Some information should not appear upon the base plan—for example, contour lines that may interfere with subsequent dimensioning or planting information; excessive texturing or delineation that may interfere later with the clarity of notations, dimensioning, and contour lines; and words or numbers that will vary from sheet to sheet. On the other hand, information missing from the base plan will have to be drawn on each sepia of the set.

Additional time may be saved by planning two or more sequences of reproduction within a set of sepias. For example, basic information is drawn on the original transparency and several sepias reproduced (first generation) as required. A second stage returns the original drawing to the drafting board and new information is added to the transparency. Sepias are then reproduced from this base informa-

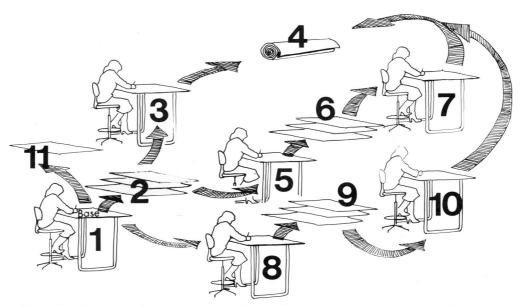

Figure 2.4 Diagrammatic representation of optional routines in preparing a set of contract documents.

tion. Thus a second drafting might add existing and proposed contour lines for use as proposed utility drawings. Only those sepias to be used as utility plans will then have contour lines. An alternative method is to add certain information on one of the first-generation sepias and then run several second-generation sepias. However, line quality of the second-generation sepia is usually diminished when compared with the first generation.

Several problems may arise when paper sepia reproducibles are used in the described manner. The quality of each sepia must be of the highest order. Each of the sepias becomes, in effect, an original base upon which contractual information is drafted. A poor-quality sepia with excessive background will be difficult to draft upon and subsequently will reproduce poor-quality prints. Poor-quality sepias will reduce the drawing's clarity and legibility. Sepia paper tends to wrinkle easily and the wrinkles will reproduce on each print. Liquid bleaches, used as erasures, will turn the sepia paper white. If a sepia reproducible has a dark background, the bleached area will reproduce white on a print, which can be a quality problem. Drafters vary in their opinion of the difficulty of drawing directly on sepia paper. Some drafters tend to smear pencil easily and generally dislike using the material.

2.7.2 Film

Polyester films, commonly referred to as plastic films, inspire many stories of their use, and misuse, for drafting. Whether an office will use the material voluntarily or is forced to use it by circumstances is not considered an issue for discussion here.

Photosensitive or diazo-sensitive film can be used in place of the sepia reproducible material previously mentioned. The drafting system is the same in both instances; that is, reproduction is made, but on film instead of sepia paper. Although the basic system remains the same, the drafting techniques, erasures, and drafting tools vary considerably between the two materials.

Film does seem to solve the problem of poor-quality reproduction often attributable to sepia. With film, all originals of a set of drawings are reproduced using the same basic material. Blue line or black line prints are generally of higher quality because of the similar base materials, and perhaps, the tendency to use ink when working with films.

Because film is easier to see through than sepia paper, its use in overlay modes is necessary to the system. Various subjects may be overlaid physically as drafting proceeds. As successive decisions are made as to the location of specific elements, these are placed over those seen below. For example, once the locations of underground utilities are determined, a planting plan can be prepared as an overlay. Such a procedure allows plants to be located in positions that will not conflict with the utility systems. Conversely, there is often sufficient reason for a utility plan to be prepared that will locate utilities away from both existing and proposed plant materials. The transparency of films allows such activity without the necessity for a light table.

2.7.3 Film and Pin Bar Systems

A special system that is closely allied to the ordinary overlaying of drawings is called a pin bar system. The feature that distinguishes this system from others is the use of small pins attached to a metal or plastic bar. Each plastic sheet of film is punched with holes corresponding to the pins' sizes and spacing. Usually the holes are punched along the top margin of the sheet. The pin bar is attached to the drafting table, and as each sheet is drafted, its registration is maintained with the pins.

The fact that each sheet is drafted in registration with all other sheets allows transparent sheets to be photographed as a group. Any sheet may be combined with any other sheet, registered by a pin bar under a camera, backlighted through the transparent materials, and a composite photograph taken. The composite photograph can then become one sheet of the contract documents. Each of the original film sheets quite often is useless as an independent bit of information. One sheet may contain only a diagram of the irrigation system, a second base sheet may contain only the plot plan information, and a title sheet might contain border lines, title block, and north sign. But when a composite photograph is taken of all three drawings, the result is an irrigation plan suitable for contracting the irrigation system.

Efficiency of drafting and the corresponding savings in time are major advantages of the pin bar system. However, any economic advantage must be measured against the costs of photo reproduction, materials, and any disruption of traditional drafting procedures.

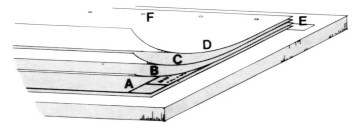

Figure 2.5   An example of pin bar drafting sequence and combinations. A—Base information limited to information necessary to each sheet in the working drawing set, borders, title block partially complete, uniform legends and symbols. B—Base site or plot plan of property lines, buildings, walks, trees, and utilities, as "existing" information. C, D—Specific subject map drawn on top of base information. E—Pin bar on table to maintain registration of drawings during drafting operations. F—Holes through each sheet. Upon completion, each final sheet will be photographed as a product of a combination of sheets A and B and C or D, etc.

2.7.4   Photography and Photographics

Today's technology allows photosensitive emulsions to be applied to papers and plastic films. Photographs may be enlarged, reduced, or contact printed directly on transparent media. Ink or pencil may then be used on the transparent medium and diazo printing obtained. Although the process is not in normal use in all offices, its application in some situations seems to have a great deal of merit.

Difficulty in relating existing conditions to those described in the contract documents and construction contracts has been mentioned. In many respects, the use of photography may be an aid to assessing existing conditions and describing the nature and extent of a contractor's responsibilities. Quite often, the use of unit prices and force account formats would be unnecessary if survey work was sufficient to help in identifying the exact nature of the work to be done.

A site project, for example, might call for the creative pruning of many mature, existing trees. Such a job is usually left for an owner's work force, requires on-site supervision by a landscape architect, is ignored due to the difficulty of verbally and graphically explaining what is desirable, or simply becomes a source of conflict between contractor and designer. Photography can solve this problem by providing a quick and very precise translation of the work description, a record of the work, and a means of measuring the work's successful completion.

If photographs of the trees are taken on transparent film, for example, they can be directly "pasted" on transparent paper, a sepia reproducible made, and the designer's notes and instructions delineated on the sepia reproducible. Prints are made of the sepia and distributed to each contractor. Each contractor can accurately assess the nature of the work and bid accordingly. Several methods exist for this document preparation and reproduction:

1.   Large-format, instant-developing camera films can be used that produce a "positive" black-and-white instant print on transparent films (available from educational supply houses for normal use with overhead projection).

2.   Black-and-white 35mm negatives may be utilized that are then darkroom developed as positives on transparent films. An enlarger can be used to "blow up" the negative in the same fashion as ordinary print enlargements on paper, except that film, such as Kodak© Graphic Arts Film, is substituted for the paper.

3.   Some paper copy machines will copy positive photos on transparent and "sticky back" polyester films that can be assembled on transparent paper and a sepia run for drafting medium.

4.   Most major reproduction companies offer a photo reproduction arm that will assist in the procedures and reproduction techniques necessary to an efficient and large-scale use of photographs and photographic drafting.

This short explanation is not intended to show photography as a panacea for site work. However, it can be valuable as a survey of existing conditions to a contract when such a survey would not ordinarily be undertaken because of cost and complexity. It is not too difficult to list a few photographic possibilities for site work problems.

1. Development of a planting plan for mature plants to be located in an existing entry court or interior planter. Design executed on an eye-level photograph.
2. Photograph of an existing plaza that indicates such features as demolition; removal, relocation, and addition of plants; lighting, benches, and paving.
3. Photograph of an eroded slope with directions as to renovation grading and replanting.
4. A photograph that describes a contractor's responsibility accurately to reproduce a historic fence or wall that no longer exists.
5. A series of oblique aerial and ground photographs that delineate the extent and nature of work on a site of dense foliage ordinarily too complex to survey.
6. A series of ground-level photographs that delineate the grading and drainage work to be undertaken as a renovation of an existing and imperfect site development.
7. Detail photographs of equipment, plant characteristics, rocks, textures, craftsmanship, and so on, to be included in technical specifications as replacements for traditional descriptive written words and phrases.

2.7.5 Overlays and Site Planning

Whatever technique is used by a drafter, it should be one that recognizes the three-dimensional character of subsurface and surface construction and installation. Site plan documents are somewhat analogous to the multiple floors of a building, that is, a vertical relationship exists among the various components.

While documents are in preparation, each subject should be related to both existing and proposed subsurface and surface features in order to avoid conflict. Although most conditions to a contract require that a designer decide the ultimate location of conflicting work, it is much easier for all if potential conflicts can be discovered on the drawing board rather than in the field.

2.7.6 Ink and Pencil

A generation gap is never more evident than in the debate normally accompanying a discussion of ink versus pencil. The older generation seems to understand pencil, the subtleness of the various gray values, the ease of erasure, and the fact that it is simply a traditional way of doing things. The debate is of particular interest as it affects working drawings. Such drawings are forever changing, and with pencil, erasing is a simple and straightforward process. However, with the advent of fast-drying inks and new types of pens, much of the frustration associated with older methods of "inking" has been removed. Anyone who has successfully mastered the inking of a drawing with a bow pen can be very proud of the accomplishment. There is a certain satisfaction in sharpening a bow pen nib, laying on the ink for that very special drawing, and not having ink all over you and the work.

Several things have happened during the past several years that seem to have changed the argument in favor of ink for contract document drawings. Important advancements have occurred in pens and drafting media. For example, new technical pens allow even the novice to lay down a straight and uniform line with line widths limited only by the number of penpoints in a set. Modern polyester films have also given both the expert and novice a chance to erase mistakes or revise at will without a great deal of damage to the film surface or drafting pride. Most technical pens are relatively resistant to drying and stoppage of ink flow. Newer inks dry quickly enough for drafting to be efficient and progressive.

However, the case for ink has really been made in newer means of production. For example, high-contrast photographic work, as with pin bar systems, simply will not work well unless ink is used for a clear and sharp line contrast. Pencil functions primarily in subtle shades of gray that photographic techniques will not reproduce. Thus inked lines are necessary to any drafting system that relies on line visibility through multiple layers of drawings or reproduction through multilayers of film or paper.

The advanced and novice drafter both should understand another aspect of pencil versus ink. The use of pencil, as mentioned, results in shades of gray whereas ink produces a totally opaque line. Thus pencil functions by contrasting shades of gray while ink must depend upon the width of a line for contrast.

Polyester films, with their different drafting approach, have created a degree of havoc for some drafters. Standard pencil leads do not work well on polyester films. Plastic leads are manufactured for line work on polyester but are often a problem because of breakage, difficulty in sharpening, and their greasy feel. Standard pencil erasures and electric erasures wear away the matte surface of films and special erasures are always necessary.

2.7.7   Hierarchy of Line Variation

The purpose of variation of line weights or widths is to increase the clarity and legibility of a drawing. When all lines, as symbols, are drawn at the same pencil weight or ink width, it is difficult for the eye to distinguish among dimensions, object edges, notations, and textures.

Although rules exist and legends abound with directives regarding proper line weights and widths, they all boil down to one thing: the constructed physical object must be clearly read along with attendant information. For example, the physical edges of objects, say a retaining wall, might be drawn with H lead or number 3 pen and the attendant dimensions and material texturing with a sharp 2/3H lead or a number 0 pen. Lines that denote a material (say, the steel within a retaining wall) might be darker than the attendant information but lighter than the object's edges. In essence, let the contractor be aware of exactly what must be done and keep attendant information subordinate and out of the way.

2.8   COORDINATION OF DRAWINGS

A major portion of document coordination will involve the prevention of overlapping, redundant, and conflicting information among drawings and technical specifications. Site work may be particularly susceptible to these problems whenever all three design disciplines are involved in a project's development.

2.8.1   Condition I

As a consultant, a designer works directly through a prime designer. In all likelihood, both drawings and specifications will assume a graphic and organizational format dictated by the prime designer. Drawings, for example, must be numbered, titled, and located in the prime designer's contract document set. The coordination of information, drafting characteristics, drawing media, lettering style, scales, indexing, title block, base map information, and those items contained in the various contracts must be carefully coordinated.

Before a consultant can begin preparing contract documents for site work, the following information should be made available.

1. A site survey of existing conditions should be prepared.
2. Exact identification should be made of those items of site work that may be delineated on architectural or engineering drawings. For example, some architectural details may set typical paving design immediately outside the building's wall, a potential conflict with redundant site drawings.
3. Location of water, electrical, and sewage facilities in the building or at the building's perimeter should be indicated. If utilities are a part of the site work, their points of entrance into the building must be known. Quite often, the building contains electric power sources, water sources, and sewage necessary for exterior lighting, irrigation, and facilities.
4. As large projects seem subject to a continuous shuffling of building locations, a fixed plan of architectural features is necessary to the location of paths, roads, drives, drainage, irrigation, and plantings.

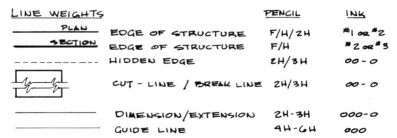

Figure 2.6   Examples of lines produced by various grades of pencils or pens. Line weight aids in a drawing's legibility.

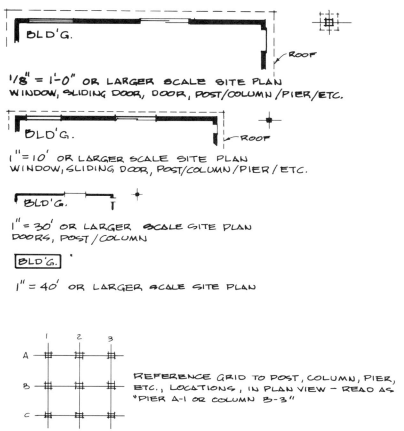

Figure 2.7 An example of line work and degrees of information obtainable by various drawing scales. Note that a relatively small reduction in plan scale produces a drastic reduction in available plan information. A grid locates elements by intersecting lines.

5. If such subjects as grading and drainage are delineated on engineering drawings, these must be made available prior to planting or walk design because horticultural environments are often drastically altered by drainage patterns as well as by soils and slope attitude. Irrigation systems also are directly related to soils, plant types, solar exposures on slopes, and drainage patterns.

6. Architectural lighting must be known in order to prevent redundancy or conflicts with exterior lighting.

7. Sizes, confirmation, and depths of wall footings of architectural or engineering design must be known in order to avoid conflicts with plantings and underground utilities.

8. Cantilevered architectural features may need plantings and irrigation beneath the overhangs that will require interpretation.

9. Architectural materials and fenestration must be known in order exactly to locate and identify such items as exterior materials, furnishings, and planting.

10. Architectural dimensions must be known in order to coordinate a building's location on the site and the location of exterior work.

11. If locations of utilities are to be handled on engineering drawings, the locations of manhole covers, clean-outs, drain inlets, telephone pull boxes (exterior), transformers at ground level, telephone and other poles, trash pick-up points, street light standards, rights-of-way, and easements must be known in order to coordinate with plantings, irrigation, walks, general paving, and similar features.

It is not considered the prime designer's responsibility (or that of the subdesigner) to separate or segregate the work of each subcontractor who might function under the prime contractor. A prime contractor will segregate as befits the situation and types of work. From the standpoint of a consultant's work, there is no great need to separate the various parts of the work into separate subject drawings or specifications. However, it is a courtesy to do so. One problem that must be considered by a consultant is the provision of information drawings and technical specifications that complement the work of the prime designer.

2.8.2 Condition II

The segregation of prime designer's documents into several prime construction contracts increases complexity for a consultant. The character and nature of each prime contractor must be easily and accurately separated in separate contractual agreements. If, for example, site work is separated into a single contract that will involve both rough grading and finish work, along with other prime contracts for foundation work, architectural construction, and utilities, the project documents and coordination of construction become complicated. While under condition I a redundancy among drawings was simply an annoyance, a redundancy among the several contracts is another thing. For example, both the site work drawings and the architectural drawings might set the finish grade for drainage. The redundancy will now be in two different contracts, with each prime contractor assuming responsibility for the work. A very common condition serves as another example. The architectural drawings carry details of planter boxes and the landscape drawings delineate plants in the planters. Unfortunately, each designer might assume the infill of soil to be in the other person's drawings and specifications.

Condition II can occur both with a prime designer and with a subcontracting designer. A designer's responsibilities to the prime contractors are the same in either event, that is, segregation and identification of the scope of work in each of the contractual agreements without redundancy. A designer's responsibilities and coordination will be very much the same among other designers but the on-site obligations will vary in the prime–subcontractor design agreement.

2.9 NOTES REGARDING DRAWINGS

In general, site work drawings will follow a traditional graphic format set by architectural drawings.

Graphic techniques useful to civil engineering are not usually identifiable with landscape architectural drawing or presentation techniques. For example, landscape architectural working drawings often depend upon contour lines to symbolize finish grading whereas engineering drawings rely on spot elevations. Engineering contracts are often based on a unit price agreement with only an engineer's estimate of quantities to guide the scope of earthwork. Data therefore are presented graphically as spot elevations with only general locations of geometrically derived slope proportions. In other words, drawings of earthwork prepared by civil engineers match their traditional system of office and design procedures, and these are often not the same as landscape architectural practices.

With this in mind, most drafters can follow architectural drawing format with reasonable assurance that it will match that of landscape architecture. Major differences are considered in subsequent chapters.

Some special features that relate to site plans are discussed in the following sections.

2.9.1 Sheet Size

Many sites are too large to fit comfortably on small sheets yet a major mistake on a designer's part is to prepare working drawings on sheets that are excessively large. Sheets more than 24 × 36 inches in size are too large to handle in the field. A site plan is often the first drawing on the site, and subject to weathering, wind, rain, and dust, and to being folded to fit into the contractor's or surveyor's pocket. Small sheets allow a contractor to review sheets incrementally as necessary to portions of the project under construction. Sheets too large reveal a lot of information that may not be needed at that particular time in construction.

An alternative is to provide the contractor with half-size drawings of the entire contract set for quick reference in the field, with the full-size set kept in a construction shack for more detailed reference. These reduced-size drawings can be produced photographically in most reproduction shops but are quite costly when only a small number of sets is required.

2.9.2 Match Lines

When a sheet size is too small to show an entire site, it is best not to reduce the site scale to less than 1 inch = 20 feet simply to fit on a sheet. The site can be divided into sections and drawn on several sheets.

However, with this procedure it is difficult for both the designer and contractor to maintain continuity in the overall picture of the project. Stories abound about designers and contractors forgetting that another sheet exists, thus producing errors in cost estimates, material ordering, and construction, and, in general, in things that just don't fit together.

When forced to use more than one sheet for a site, several rules should govern drafting.

1. Place a match line at a point that can be located on the site, preferably at a point that is easily and visibly identifiable in the field. Avoid match lines that only the drafter can see or that require on-site survey work to locate.

2. Locate match lines so that they can be found by dimension, if necessary. For example, a match line can be located by surveying dimensions off property lines, at stations along a road, or at grid intersections (see Chapter 5 for clarification of terms).

3. Prepare a "key" map of the entire property immediately after the site's increments are determined. Double check the small-scale site plan with each sheet to make sure that the entire site is covered by all of the drawings. Do not take a chance of developing an entire set of drawings only to discover that a piece of the property is missing. The key map should be a part of the contract documents so that a contractor can relate each separate sheet to the whole of the property and scope of work.

4. Do not draft any contract work outside of the match line. The practice of drafting outside of a match line may be defended on the premise that it assists the contractor in orienting to other parts of the property. This may be true. However, what usually happens is that work items, quantities, and the like are counted twice—that is, are counted on one sheet outside the match line and within the match line on another.

2.9.3 Diagrammatic Presentation

A majority of site work drawings will be diagrammatic in their presentation. A diagrammatic drawing will generally delineate patterns, arrangements, and parts in graphic outline form. Site plans are usually of such small scale that not every bend and twist of a pipe or wire, either vertical or horizontal, can be shown in exact form. In point of fact, a contractor must be able to develop the final location of components during the construction and installation phases. A diagrammatic drawing simply establishes the character and scope of the work through the graphic use of symbols. It is imperative that a drafter clearly delineate a system and identify its components, regardless of the real location of materials. A contractor needs to see what connects to what, the size of materials, and the overall pattern as a diagram.

Many technial specification or contract conditions will clarify the diagrammatic nature of specific drawings by a paragraph similar to the following.

Because of the scale of the drawings, it is not possible to indicate all offsets, fittings, and appurtenances that may be required. The contractor shall carefully investigate the structural and finished conditions affecting all contracted work, and plan the work accordingly, furnishing such fittings and appurtenances as may be required to meet such conditions without conflict. Drawings are generally diagrammatic and indicative of the work to be installed.

The use of such a paragraph allows a contractor the flexibility necessary to route or locate components with respect to detail construction, locations of existing trees, underground obstructions, and so on, while the drawing controls the overall pattern of the work.

It is extremely important that a drafter strive for clarity in presentation. Electrical wires, for example, need not be drawn exactly where they will be installed, but it is important that the general direction and connections to the various components be shown. Pipes should be drawn with sufficient graphic space between them that sizes can also be shown, that one pipe not be confused with another, and that it is clear as to which pipes are connected or are not connected together.

2.9.4 Representational Presentation

A set of working drawings is generally not the place for representational drawings when such graphics confuse contractual information. Particular concern must be given to planting plans that attempt to portray the future size and texture of plant materials. Such drawings not only are time consuming to prepare, but may simply prove to be graphically "busy," of little value to a contractor, and unclear.

2.9.5 Indexing

Whether a set of drawings is composed of one or 100 drawings, there must be some way to locate information among the various sheets. Such a system is called indexing or, occasionally, keying. A system will begin with the identification of each sheet by, usually, a number. Two things happen here. First, the individual sheet is identified for legal purposes; second, each sheet is identified so that reference can be made to that specific sheet. This may appear to state the obvious, but the courts and arbitrators are constantly involved in cases that question exactly which drawings and how many constitute the true scope of an agreement. It takes little imagination to foresee mistakes in collating sets of drawings and overlooking sheets, or a contractor claiming that a sheet is missing.

Each numbered sheet is identified in an index as part of the project manual and agreement between the owner and contractor. If necessary, and it often is, each sheet should carry a date of issue to avoid confusion with office or field changes that alter the scope of work borne by that particular sheet. Sheet numbers are generally composed of two numbers. The first identifies that particular sheet and the second denotes the total number of drawings in the set. For example, "sheet 2 of 5" indicates that five sheets are in the set and the particular sheet is number 2. If the construction agreement was signed 2/5/80 and a revision to the drawing components was made on 2/25/80, it is obvious that such a revision must have been accompanied by a change order.

Particular attention must be given to conditions I and II noted in Section 2.8. Site work drawings may be located, for example, in a set containing architectural or engineering work and must be identified and indexed within the same system. Under these conditions, an irrigation system drawing might carry the designation IS-2/5, indicating that it is an irrigation system and not a fire sprinkler system and the sheet is the second of five. In this instance, the five IS sheets are listed as *inclusive* in the specification manual index and therefore as a portion of the total number of drawings composing the contract document set.

The sheet identification number should be located in the lower right-hand corner of each sheet for the convenience of everyone. When many sheets of drawings are laid out for review, it is simply a courtesy and traditional for the sheet number to be legible in that corner of the drawing.

How many symbols are necessary is a matter of judgment. In general, it is a common mistake to use too few index symbols. It often helps to assess both the quality of the line weight and the number of index symbols. For instance, if the drafted line weight itself helps to delineate the edge of paving, perhaps one index symbol will be sufficient to refer to the proper detail. However, if the lines are all of similar weight, it may be difficult for a contractor to distinguish among edges. Also, one index symbol on one sheet does not carry easily over to another, nor will one index symbol usually suffice for elements that occur over an entire sheet.

Several indexing systems exist that are keyed to the sheet identification system. For most site plans, an indexing system is critical because of the number of times various details and construction activity refer to the site plan. See Section 4.4 for additional information.

2.10   DRAWING AND SPECIFICATIONS

Although drawings and technical specifications are contractually one document, much misunderstanding and many errors are traceable to the lack of a

 - SYMBOL PLACED ON SITE PLAN OR DETAIL OR BOTH DENOTING A REVISION MADE TO A CONTRACTUAL ITEM EITHER BEFORE OR AFTER AN AGREEMENT. FOR EXAMPLE :

ITEM	DESCRIPTION	DATE	ADDEU.	CH.O.
1	ALTERNATE #1 ADDED BY L.A.	1-3-82	#2	
2	PLAY EQUIPMENT DELETED BY OWNER	1-6-82	#3	
3	CONC. WALK ADDED BY OWNER	2-25-82		#1

Figure 2.8   An example "revision schedule" placed on a drawing.

clear distinction between the two types of documents.

The most common error usually occurs on the drawing board. This error is redundancy. Drafters who mistrust technical specifications may attempt to explain everything on the drawings without coordination among the written specifications. It is altogether too easy to sit at a drafting board and dream of problems or conditions of which the contractor must be made aware. A common solution is to try for exactness through a profusion of hand-lettered notations and written directives. If these notations are redundant or at odds with the technical specifications (and they often are), a contractor is left with confusion.

In addition to the waste of drafting time and the confusion inherent in redundant information, it becomes next to impossible to locate and revise redundant information when a revision is necessary. For instance, a drafter mentions the type of wood necessary to construct a wooden deck on three drawings and in several sections of the technical specifications. The owner then changes the program to a different type of wood. Hours can be wasted in trying to find all the places where the notation might appear in the drawings and in the specifications. One way to avoid the overzealous provision of redundant information is to imagine the contractor calling you each evening to request clarification or change orders.

At issue here is the basic fact that drawings very often assume a role that goes beyond their primary dependence on graphic information. Whenever notations and words become excessive, drawing information will often conflict with the specifications. In order for the drawing to complement specifications, a drafter must understand the contents and purposes of written specifications.

SELECTED READINGS AND REFERENCES

Giummo, Vince. *Overlay Drafting Techniques*. Rochester, New York: Eastman Kodak Co., 1978.

Kerr, Kathleen W., et al. *Cost Data for Landscape Construction*. Distributed by *Landscape Architecture Magazine*, Louisville, Kentucky.

Lewis, Jack R. *Construction Specifications*. Englewood Cliffs, New Jersey: Prentice-Hall, 1975.

Marsh, Warner L. *Landscape Vocabulary*. Los Angeles, California: Miramar, 1964.

McHugh, Robert C. *Working Drawing Handbook*. New York: Von Nostrand Reinhold, 1977.

Stitt, Fred A. *Systems Drafting, Creative Reprographics for Architects and Engineers*. New York: McGraw-Hill, 1980.

Weis, Hermann W. *Landscape Architectural Construction Detail Drawings*. Athens, Georgia: University of Georgia Press, 1979.

Chapter Three
Existing Site Conditions

*Knowledge of a site's existing conditions, necessary to design and contracting,
must be precise, accurate, quantifiable, and clearly communicated;
anything short of such quality is usually what you receive.*

John M. Roberts

Existing site conditions include the location and characteristics of various physical and legal features that exist at the time a contractor begins work. From the standpoint of a designer, work on a design concept may not begin until at least basic information is received. Contract document preparation may not begin until detailed and specific information is received.

Two premises must be recognized by a designer and owner. The first concerns whether or not survey information given or reviewed by a contractor is to be a part of the agreement. A second premise concerns whether or not information is guaranteed or warranted as to accuracy or is simply an estimate of probable conditions. Both owner and contractor must have a clear understanding of these two premises prior to bidding and offering or accepting an agreement.

A majority of disputes may be directly linked to a misunderstanding as to the validity of information regarding existing site characteristics. Generally such misunderstandings may be traced to the language of the documents. Many cases involve the question of a contractor's fair assumption of risk during bidding and implementation, that is, whether both parties fairly understood their respective assumptions of risk and the potential economic loss from delay or excessive direct costs.

3.1 THE DRAWING AS INFORMATION

A drawing depicting existing conditions is variably referred to as a plot plan, site plan, existing condi-

tion plan, site survey plan, demolition plan, or site preparation plan. No doubt the number of titles stems from regional traditions and whether or not a drawing contains proposed work in addition to existing conditions. Drawings portraying only existing information are referred to as a plot plan, existing condition plan, or site survey plan. Other titles are indicative of special information. For example, a site plan might contain existing site information, demolition directives, and proposed improvements on one sheet. A demolition and site preparation plan will contain existing conditions as well as directives for major demolition, relocations, cutting and patching, waste removal, or other work to be completed prior to new construction.

3.1.1 Contractual Information

Existing conditions will include, but are not limited to, property lines, easements, topography, buildings, paving, curbing, types of vegetation, site features such as rock outcroppings or water bodies, subsurface and above-ground utilities, off-site utilities, storm water and sewage connections, north direction, vertical data, rights-of-way, streets and roads, walks and trails, walls, fences, soil characteristics, legal descriptions, and geology. Although many of these features are normal to surveying practices, it may be necessary for a designer to assist the owner in precisely defining information needs.

Information included on a drawing or bound in a project manual is considered part of the contract agreement unless stiplulated otherwise. Informa-

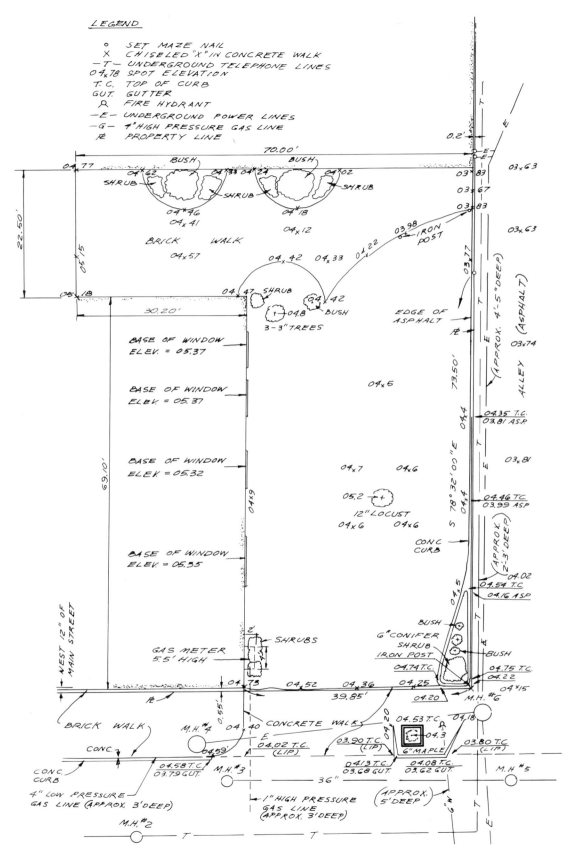

Figure 3.1 Example of a standard survey map prepared by a registered surveyor.

tion that appears on drawings or specifications carries with it an implied warranty of its accuracy. A contractor is entitled to presume that such documents contain all available information and data regarding a project. In general terms, information will be considered as fairly representing a description of existing conditions unless *expressly disclaimed.*

Attempts to shift the burden of information guarantee from the owner to a contractor usually occur as a *disclaimer* provision in the general conditions of the contract. The *General Conditions of the Trustees of the California Colleges* (July 1965 issue) may be a useful example of such a provision's language:

> *Where investigations of subsurface conditions have been made by or on behalf of the Trustees in respect to foundation or other structural design, and that information is shown in the plans, said information represents only the statement by or on behalf of the Trustees as to the character of material which has been actually encountered by it in its investigation, and is only included for the convenience of bidders.*
>
> *Investigations of subsurface conditions are made for the purpose of design, and neither the Trustees nor the Architect assume any responsibility whatever in respect to the sufficiency or accuracy of borings or of the log of test borings or other preliminary investigations, or of the interpretation thereof, and there is no warranty, either expressed or implied, that the conditions indicated are representative of those existing throughout the work, or any part of it, or that unlooked for developments may not occur. Making such information available to bidders is not to be construed in any way as a waiver of the provisions [respecting contractor's personal site investigation and additional compensation] and bidders must satisfy themselves through their own investigations as to conditions to be encountered.*
> [Language within brackets reconstructed by author.]

Even with such a disclaimer, several issues will remain to cloud the interpretation of existing information. A contractor's ordinarily limited opportunity to thoroughly investigate site characteristics may be an issue even though each contractor is directed to visit the site. Both owner and designer are considered to have relatively superior knowledge of the site and its characteristics as compared with the limited opportunity given to a contractor during a bidding period. A site visit may not in fact bar a contractor's recovery of extra compensation when it can be proved that the nature of existing conditions was unforeseen due to variance with

normal regional conditions and that reasonable efforts by the contractor did not discover unusual conditions.

In addition, owner-provided information may vary excessively from reality and produce a hardship regarding a contractor's method of excavation or equipment requirements. A contractor may bring action if, for example, factual information is misleading, information is knowingly withheld by the owner or owner's agents, or a contractor is somehow prevented from discovering relevant information.

Another issue concerns the implication of accuracy or warranting of the information given to a contractor. If, for example, it is proved that information was invalid or fraudulent, a contractor may seek extra compensation. A lack of accuracy may extend to the manner of determining, for instance, the character of soils or the location of old concrete below grade, which may be incorrect owing to human error or incompetent investigation.

Embodied in existing condition information is a promise from an owner to a contractor concerning a site's condition, and perhaps a date on which to begin work. Generally these issues will involve a delay to the contractor if the owner does not have the site available and in the condition described and delineated in the contract documents. Under multiple contract situations, delay is often caused one contractor by another contractor's failure to finish on time. While a late contractor may have been the root cause for delay, it is the owner's ultimate responsibility to make the site available to the contractor on time and in the condition delineated by the agreement.

Particular care must be exercised when identifying existing conditions on a drawing. A drawing is not the place on which to interpolate vertical elevations, locate elements by memory, or otherwise add to existing information in a nonfactual manner. On those occasions when existing conditions are approximate, they must be identified by notation, that is, the contractor must be informed that portions of the site information may be approximate and require verification. Special concern must be given to site features that will tie directly into new construction or installation work. Due care and skill will involve the contractor's ability to price the work with full knowledge of what is and what is not known about existing conditions at the site.

Issues involving existing conditions are never completely clear or without contingencies in site work construction. Resolution of conflicts is part of the construction experience. Contractors are sel-

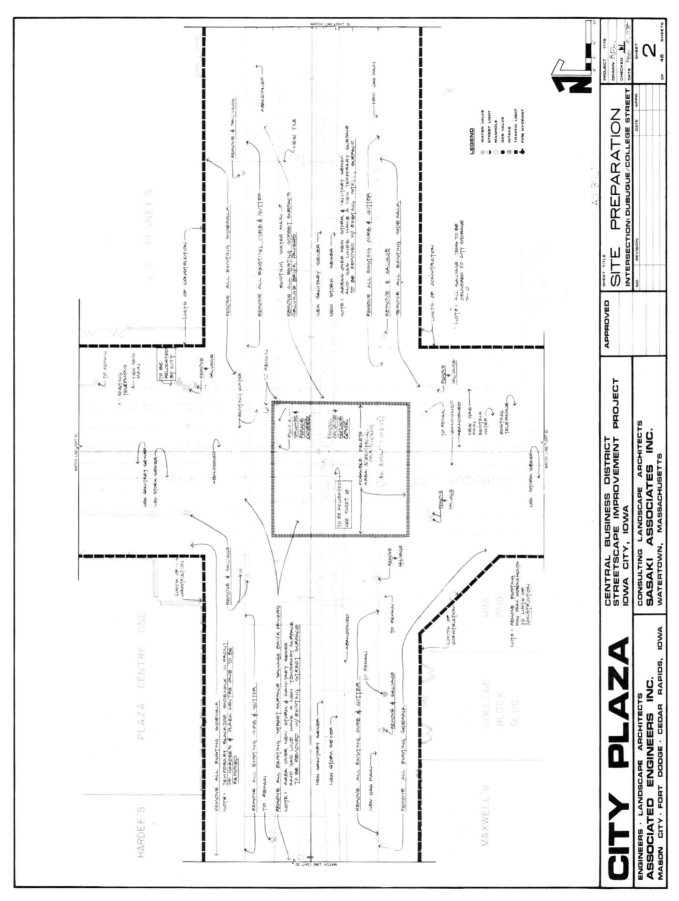

42

dom surprised by what they may find. At issue is who pays for the surprise.

3.1.2 Noncontractual Information

It is common practice for certain existing site information not to be bound with the project manual and to be purposefully excluded from an agreement between owner and contractor. For example, soil tests, soil boring data, and other information pertinent to the designer and the design process, or an owner's inventory and survey of adjacent property conditions, may be held on file. Such information and its locations should be inventoried and noted in the contract documents and be open to review by a contractor prior to acceptance of any bids. In other words, a contractor may be required to perform a personal site visit plus a review of whatever information may be on file and available to a designer or owner.

Exclusion of information from contract documents by a failure to reveal does not necessarily bar a contractor from recovery of extra funds for extra work. If, for instance, a contractor experiences extra work and can prove that certain information was available from either the owner or designer and its revelation could have prevented extra work, a contractor may recover damages from a designer or owner, or both.

3.2 RESPONSIBILITY FOR INFORMATION

Defining responsibilities for existing site information begins at the time an agreement is executed between owner and designer. An agreement must state and define responsibility, payments, and information.

3.2.1 The Owner

An owner is commonly made responsible for providing existing site information. Normally an owner will hire a registered surveyor to make boundary topog-

Figure 3.2 An example site preparation plan indicating that certain items must be removed, abandoned, repositioned, or protected by the contractor prior actually to beginning new construction. Note the "limits of construction" line that legally limits the maximum extent of both demolition and work. (Courtesy Jack Leaman, landscape architect, Ames, Iowa)

raphic and physiographic surveys. The surveyor will deliver a finished land and site survey to the designer.

An owner essentially has two options. The first option is to retain, acquire, and obtain all information possible where knowledge, conditions, and instruments allow measurement. This option will reduce the risk to a contractor and increase the owner's costs to survey. A contractor's risk is even further reduced by inclusion of all information in the contract documents and an owner's warranty of the information's accuracy and reliability. Of course, this may also be coupled with a compensation arrangement that further reduces the contractor's risk. However, in practice an owner can usually gain only a portion of the information possible. A second option could be, in the extreme, *not* to provide any information regarding the existing site conditions. Under this option, a bidding contractor must make the investigation and establish the degree of acceptable risk. The owner simply does not accept responsibility for information about the site, and therefore accepts the economic contingencies inherent in construction bids and a design based on high-risk factors.

3.2.2 The Designer

Traditionally, after a surveyor delivers a site plan survey to a designer, work will commence on the various phases of the concept and its implementation. A surveyor's plan often might be traced and become the basis for the existing condition information.

A degree of responsibility might be assumed by a designer, for example, when survey information is traced, gathered, or otherwise interpreted from data supplied directly by a surveyor or consultant. For instance, suppose an owner supplies a written legal description of the property and a graphic interpretation is made by a drafter. An owner might retain a soil consultant to test horticultural soil factors and this information might be translated into detailed technical specifications. In both instances, the professional designer functions as a translator and accepts at least partial responsibility for accuracy. Many designers include a copy of a surveyor's survey with their set of drawings as a part of the contract documents and also list acquired data or consultants' information as part of noncontractual information for courtesy review by the contractor and to mitigate responsibility.

It is not perfectly clear that a designer may avoid responsibility for the accuracy of information even when so stated in an agreement. It is best that a designer presume that the courts will assume a professional is in complete control of a situation and that an owner is at a disadvantage. For example, if an owner unknowingly supplies out-of-date, incorrect, or otherwise imperfect information, it may well fall to the professional to recognize the problem. Most owner–designer agreements make the owner responsible for provision of information, however, a designer's actions or lack of action may negate such responsibility.

3.2.3 Concealed Conditions

Allowing for negotiation of extra costs incurred by the contractor whenever unforeseen and concealed conditions surface requires special contractual language. Extra cost recovery reduces a contractor's contingency. Federal construction contracts and AIA document A201 both contain a clause allowing for equitable adjustment of costs due to concealed conditions. Such a procedure shifts the burden of risk for concealed conditions from a contractor to the owner. It should be noted that the inclusion of such a clause for adjustment of costs due to concealed conditions is directly opposed to the language and intent expressed by, for example, a disclaimer as previously noted in Section 3.1.1. The *General Provisions of the General Services Administration* (October 1969) may serve as an example of language allowing for recovery of extra costs for concealed conditions:

> *The Contractor shall promptly, and before such conditions are disturbed, notify the Contracting Officer in writing of: (1) subsurface or latent physical conditions at the site differing* materially *from those indicated in this contract, or (2) unknown physical conditions at the site, of* unusual *nature, differing* materially *from those* ordinarily *encountered and generally recognized as inhering in work of the character provided for in this contract. The Contracting Officer shall promptly investigate the conditions, and if he finds that such conditions do materially so differ and cause an increase or decrease in the Contractor's cost of, or the time required for, performance of any part of the work under this contract, whether or not changed as a result of such conditions, an equitable adjustment shall be made and the contract modified in writing accordingly.* [Emphasis added]

The quoted section goes on to outline the delivery of notices, and timing of the procedures. The language of the example clause strongly indicates that the words *materially* and *unusual* will be subject to debate and negotiation as to their meaning. Obviously, if an owner's agent should find that conditions are not materially different from those shown in the contract documents, ordinarily encountered, and recognized as inhering within the work's character, the contractor cannot recover "extra" costs. However, differences of opinion will probably result in litigation or arbitration of opinions, with the burden of proof resting upon the contractor.

Contract documents that include an owner's disclaimer of responsibility or warranty of information, *and* also allow a contractor access to negotiation for extra compensation for work that is materially different, will lead to problems in sorting out the true intent of the parties. If a contractor is forced to make a complete site investigation, an allowance for concealed conditions is negated. On the one hand, an owner is saying, "Here is some information that I do not warrant." On the other hand, that owner is also saying, "If you run into trouble, I will share the risk." In either event, the terms *material* and *unusual* will be defined with respect to the language and conditions of the contract documents. It is important that existing conditions and intended risks be fully examined in the contract documents.

3.3 MULTIPLE DESIGNERS AND CONTRACTOR INTERDEPENDENCE

When multiple designers are simultaneously involved on one site development, existing conditions for one designer may depend upon the final drawings and contract performance of others. Under these conditions, the accuracy, extent, and timing of existing contractual information potentially are hampered and liable to malpractice actions, change orders, and delays. For example, a prime architect designs and a prime building contractor constructs a building with partial site preparation included in their work. The owner then contracts with a prime landscape architect for preparation of complete site work drawings to include finish grading, planting, irrigation, and exterior lighting. The site design concept and the work of a prime landscape contractor will thus be dependent upon those conditions remaining after the prime building contractor completes the contracted scope of work. The source of a landscape architect's information as to conditions that *will* exist at the time a landscape contractor begins work becomes the architect's contract documents. Failure of the building contractor to perform all contracted work or changes in the scope of work will impact upon the landscape contractor's scope of work.

When large-scale developments involve several prime designers and contractors, the cumulative effect of contract imperfections, construction errors, changes, and confusion always produces poor information. It is difficult for a single designer to change such a situation. A designer must clarify and distinguish among the fixed, semifixed, and changeable elements of information or the entire project may become chaos.

3.4   LUMP SUM AGREEMENTS

A lump sum agreement essentially accounts for all work occurring between existing and proposed conditions. If an error exists in ascertaining existing conditions, it will be significant to all parties. The nature of a lump sum agreement thus presumes contractual knowledge of existing conditions as well as the scope of proposed work.

A potential problem for all parties arises with respect to the time of an original survey and the actual time at which construction and installation begin. If existing conditions change, the degree of change will be judged relative to those existing conditions in the documents. Although a contractor is usually contractually cautioned and directed to verify existing conditions, it is not without legal precedence for major changes in existing conditions to void an agreement. For instance, an act of an owner or of nature may cause significant revision to a site's topography and change the original conditions under which a lump sum agreement was arranged, thereby voiding the agreement. However, in normal practice such problems are usually adjusted by change order to the contract.

3.5   UNIT PRICE AGREEMENTS

Unit price agreements often develop because of a lack of information about existing site conditions or the quantity and quality of work expected of a contractor. Under such contractual obligations, a

great deal of latitude will exist in defining and describing existing site conditions. However, discovery of unusual revisions to a contractor's estimated quantity of units that significantly change the scope and character of a contractor's work may be grounds to void a unit agreement or seek a change order. The definition of significant change would probably be based upon the contractor's or owner's claim that the statement or acceptance of a price would not have occurred if a better estimate of existing conditions had been available.

Both the owner and contractor may be protected by including a definition of *major* or *significant* change to the quantity of units or the unit prices in the general or special conditions to the contract. For instance, 20% or more might be considered major or significant change to the whole of the contract whereas 10% might be major or significant change to any one unit's quantity.

3.6   ACCURACY TOLERANCE—HORIZONTAL

A designer will be concerned with the degree of accuracy necessary and suitable for project and site development. Site plans will generally be useful when prepared at $\frac{1}{8}$ inch = 1 foot to 1 inch = 20 feet, with 1 inch = 30 feet as a marginal scale. A land surveyor will supply the plotted information at any scale requested by the owner or designer. The horizontal data and property description should be graphically prepared within those surveying tolerances accepted locally with property corners referenced to local coordinates, property numbers, rights-of-way, section numbers, easement locations, and other pertinent information. If exact locations of existing physical elements are necessary, the owner should request that numerical data be supplied from fixed locations. Special elements, such as trees, must always be specified as to minimal locational accuracy.

When precise property lines or similar accuracy is not necessary, aerial photography is capable of producing a horizontal accuracy that normally equals a graphic tolerance of 0.02 inch. At a scale of 1 inch = 30 feet, the scaled accuracy of mapped features would be no better than 0.6 foot whereas at 1 inch = 100 feet, it would be approximately 2.0 feet. Grid coordinates can be plotted to an accuracy of $\frac{1}{1000}$ inch or about 0.3 foot at a scale of 1 inch = 30 feet.

Both the contractor and designer may require

identification of the individual parcel or several parcels that make up the work site. All legal endeavors, such a liens, billings, ownership, and material and equipment attachments, will require property identification as a filing requirement. When a contractual site includes several lots or parcels of property, each lot must be identified for purposes of, for instance, liens. Particular attention must be given to the legal ramifications and liens involving sites that are not contiguous.

3.7 ACCURACY TOLERANCE—VERTICAL

Existing topographic and finish structure elevations can be determined by field or aerial survey methods. Small sites usually must be surveyed by instruments whereas large or complex sites may require aerial techniques.

3.7.1 Instrument Survey

Small parcels of land and those that are simple enough to allow field crews to enter and visually examine them can be surveyed topographically to tolerances potentially greater than by aerial methods. However, for ordinary work, and except for the setting of property lines, vertical surveys commonly require 100% of the spot elevations to be within ± 0.10 foot. Contour lines are usually interpolated from data surveyed at intersections of a horizontal grid—that is, spot elevations are found at the intersections of a 100-foot horizontal grid, for instance, and the contour locations estimated from such data. Greater accuracy would be possible if the grid were perhaps 50 feet rather than 100 feet. It is best to consult with a land surveyor regarding acceptable accuracy and costs relative to the owner's responsibility.

3.7.2 Aerial Survey

A decision regarding survey accuracy will reverberate through the owner's costs, design complexity, and the potential method of contracting new work. Costs of both a field and aerial survey, for example, may increase by a factor of 4 as the contour interval is halved. A contour interval that is too great will not provide sufficient information for accurate estimation of earthwork quantities, and will provide high risk to a lump sum contract.

There exists a practical limit in graphic separation between the horizontal position of contours of about 0.05 inch. At a photographic scale of 1 inch = 30 feet, contours cannot be drawn closer together than 1.5 feet, whereas at 1 inch = 100 feet, the minimum distance would scale about 5 feet.

A limit also exists as to an aerial survey's practical accuracy with respect to contour lines and spot elevations. A designer can expect that 90% of the elevations indicated by contours will have an accuracy of one half of the contour interval or better. The remaining 10% of the contours can be in error by one contour interval or less. Ninety percent of all spot elevations derived from aerial photogrammetry should be accurate to at least one fourth of the contour interval, and the remaining 10% may not be in error by more than one half of the contour interval. In other words, a contract grading plan with a 1-foot contour interval may have 90% of the existing contours in error by ± 3 inches and the other 10% by ± 6 inches.

The example given is within the photographic rule of thumb that allows an original map to be "blown up" a maximum of five times. By this rule, the original map prepared at 1 inch = 100 feet can be enlarged to 1 inch = 20 feet with a marginal loss in quality. However, at original map scales of 1 inch = 40 feet and smaller, there usually will not be sufficiently detailed site features for its use as an existing condition map or site plan. Supplemental on-site surveying must be specified for detailed feature identification. Normal planimetric aerial surveys at scales of less than 1 inch = 40 feet will not usually delineate such features as manholes, detailed building conformations, curbs, sidewalks, and hydrants, or their respective spot elevations.

3.8 UTILITIES AND DRAINAGE

Surveying various aspects of underground utilities and storm drainage structures may be likened to predicting the weather: a certain amount of scientific knowledge and instruments are available but sooner or later one has to make a guess. The degree of accuracy in surveying underground elements is related to the age of previous work, memories, and how organized available information might be. Success and accuracy in locating and identifying subsurface conditions are critical to the degree of risk shouldered by the owner and/or contractor.

If installations are relatively recent, it may be possible to refer to "as-built" or "as-is" drawings. An as-built drawing records subsurface installations as the previous contractor left them. As-is drawings record the owner's additions to or corrections of the systems during their life. Unfortunately as-is information is only occasionally available for private work, although public work almost always retains up-to-date information. As-built drawings may not truly reflect the present condition of any underground system. The value of these drawings, if they are available, should be certified by the owner and their accuracy scrutinized and brought as up to date as possible.

Metal pipes can often be located and surveyed by the use of sensitive metal detectors. If an electrical wire was run alongside nonmetal pipe during installation, it is sometimes possible for the detector to locate the metal wire or a low-level electric charge to be given to the wire during the survey period. The contractor should be made aware of the reliability of such methods, when they serve as a basis for surveying data.

Assuming that all necessary data can be made available to the designer, such data must be divided into two separate categories. One category includes the data important only to conceptualizing and directing the work to be accomplished by the contractor. A second category contains that information of importance to establishing the obligations of the contractor. For design purposes, a designer needs to know the existing pipe size, type, condition, location, pressure rating, and the manufacturer's various recommended means of connecting to the pipe. A contractor needs to know the existing pipe's location, where the designer wants the connection made, exactly what types and sizes of materials are to be used, and how the quality of workmanship will be measured (assuming the contract is not a performance-type obligation). The designer has accepted the responsibility for designing and the contractor has accepted the obligation to install as specified. (Not all contracts are handled in this manner.) If, however, the existing pipe is not where it is shown on the drawing, the contractor spends all day looking for it, and as the owner has warranted the pipe's location, a designer can expect a request for a change order.

Responsibilities are related to Section 3.1.1; that is, what does the agreement indicate to be the assumption of risks for the owner and the contractor? Does the owner disclaim, per Section 3.1.1, any responsibility for the accuracy of information or share in the risk by assuming some responsibility? Do the contract documents reflect the true intent of the two parties by distinguishing among estimates, guarantees, and a reasonable attempt at surveying, and was the subsurface underground condition materially different from that shown on the drawings? Did the owner make available all known information?

3.9 VEGETATION

Techniques, accuracy, and costs of inventorying individual plant materials will depend on their growth patterns, size and form, season, and arrangement. Sites containing trees with single trunks, arranged in a uniform pattern or spacing, and unencumbered by obstructive underbrush, can be surveyed visually by the use of a plane table or, occasionally, by high-resolution aerial photography. Normal aerial surveying agreements do not ordinarily include the mapping of individual tree locations (except isolated specimens and masses) or the identification of species. In most instances, an individual tree survey must be accomplished by ground survey with plant identification assistance from a landscape architect, botanist, arborist, or horticulturist. Aerial surveys are only approximate in their accuracy, even when the trees are deciduous and photographed in the winter.

A few projects may include a survey of vegetative health, age, and management practices. Such a survey is possible only with a ground survey crew and consulting arborist.

A site planning vegetation survey must not be confused with an ecologic survey. Such a survey will concern itself with transects and plot counts of all vegetation, that is, from grass clumps to trees. Generally such information will be a prerequisite to master planning and management plans but too finite for use in ordinary contracts for construction.

When drawing an agreement for survey work, an

Figure 3.3A *(Page 48)* Typical aerial photograph of a site prior to the development of a site survey or contour map. Physical objects are shown and an impression of ground form can be detected. (Courtesy Aerial Services, Inc., Cedar Falls, Iowa)

Figure 3.3B *(Page 49)* A Contour map of the area shown in Figure 3.3A prepared by means of a "stereo plotter." Note that both an aerial photo and topographic map may be very useful in analyzing the site features, paving, existing land use, vegetation types, and circulation paths. (Courtesy Aerial Services, Inc., Cedar Falls, Iowa)

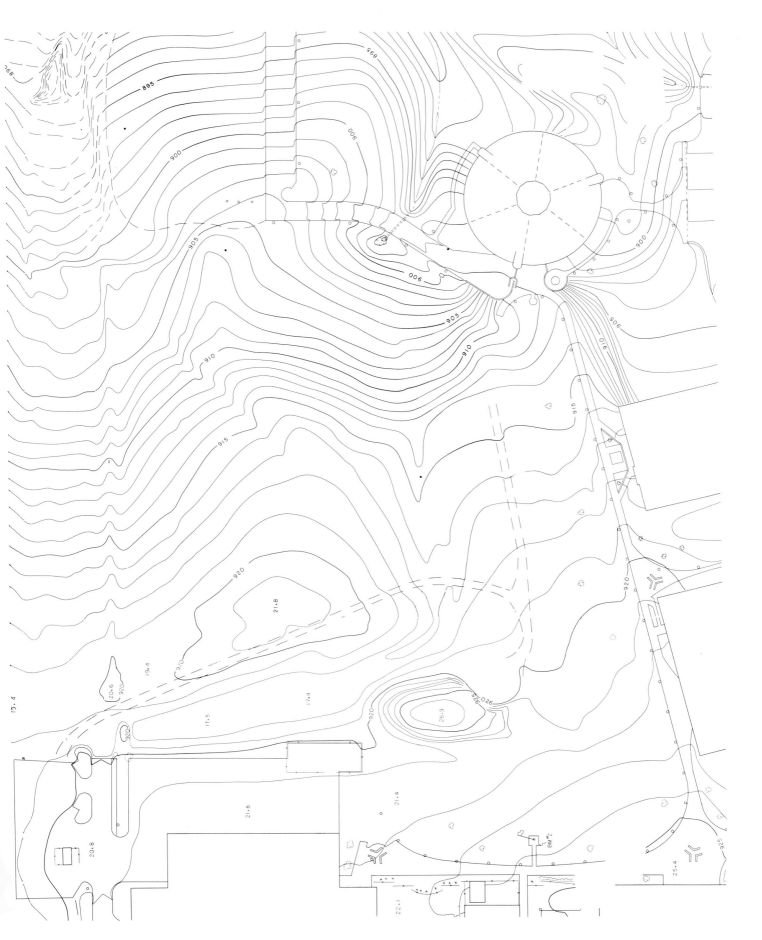

49

owner should clarify the need for genus, species, height, condition, form, and height to first branch of trees, and whether or not the diameter will be taken at soil level or at breast height (DBH). When some or all of the surveyed vegetation may require attention by the site work contractor, the survey crew should number or otherwise identify each tree with a permanent marker located consistently on, for example, the north side and at the DBH. During the survey period, each tree may require the writing of a work order type of specification by an arborist, describing the nature of the work to be accomplished, such as surgery, guying, or cabling.

Some sites may require contractual arrangements similar to those of subsurface obligations, inasmuch as the degree of risk is proportional to the amount of information obtainable by the owner.

The proper identification of plant species is always necessary to the designer. Only exact genus, species, and variety will assist the designer in determining a plant's probable reaction to, for example, earthwork, excavating, chemical management, transplanting, and maintenance management.

3.10   SAFETY

Although it may not always be clear as to the legal responsibility, it seems prudent for a designer to assume the identification of existing conditions that actually or potentially will affect the safety of people or animals. During a site survey and a design professsional's site reconnaissance, obvious conditions affecting safety at least should be described to the owner and, if possible, corrected as part of the contractual work. The designer's presence on site may lead, if problems develop, to accusations that a reasonable professional ought to have observed and should have notified the owner of conditions of hazard—for example, weak trees, excavations, displaced paving, unsafe circulation, unsafe utility conditions, noncompliance with codes, hazardous railings, or unsafe structures.

3.11   ENVIRONMENTAL IMPACT

Ordinary survey techniques do not include the specialized inventorying of environmental factors. Environmental factors, such as climatologic conditions, vegetation communities, sociologic conditions, and economic or feasibility studies, are generally considered planning and design information only indirectly associated with implementation. (See Section 6.10.)

3.12   SOIL DATA

An analysis of engineering and horticultural soil properties will usually be necessary for any site work involving the construction of buildings or other structures and the installation of plantings. Engineering properties include a soil's texture, moisture–density relationships, and structural strength. Horticulture properties include moisture conditions, fertility, tilth, organic content, and pH.

Data concerning general engineering and horticultural soil properties are obtainable from various governmental agencies that supply public information. However, most such public information is too general and the maps too large in scale to pinpoint particular sites with specific information. The U.S. Department of Agriculture's Soil Conservation Service field offices can be a good source of some horticultural data, but it must be remembered that agriculture is the Service's major occupation and that private commercial work is of only indirect concern.

In general terms, horticultural information does not become contractual, that is, included in contract documents. Seldom does a contractor find it necessary to review horticultural information. However, if a contractor's work is to be judged by (1) performance specification granting a contractor responsibility for soil amendments, (2) contractor guarantee of plant growth without conditions, or (3) contractor work that includes maintenance, there may be a need for a contractor's review of information. In other words, the entire responsibility of the contractor must be examined in light of soil information.

3.12.1   Horticultural Soil

The need for information about the horticultural properties of soils is linked to the site, the land use, and the proposed nature of plantings. If, for example, a site is to be undisturbed by a land use, presently supports native plant materials, and only native plants will be added to it, there is very little reason for any extensive soil data. On the other hand, a site that will be or has been urbanized or subjected to mass cut and filling of soil, or one to which irrigated ornamental plants are to be added, will require extensive soil investigation as to properties of the various existing soils.

It is common for contract documents to describe and specify exactly what the contractor is to apply to the soil and, generally, the incorporation of materials into the soil. The following include the type of information that may be necessary to the specifications.

pH

This is a principal indicator of a soil's acidity or alkalinity. The pH expression is the logarithm of the reciprocal of the hydrogen ion concentration (potential hydrogen) expressed on a logarithmic scale of 0 to 14, with 7 considered neutral. Soils testing pH below 4 or above 9 are considered toxic to most plants. Whether or not pH value requires consideration in a contractor's work will depend a good deal on the characteristics of the plants to be installed. For example, a pH of 8 may be just the thing for plants tolerant of alkalinity but may require extensive application of soil amendments and minerals if acid-tolerant plants will be planted as part of the contract work. Concern must be given to the pH expression in judging the acceptability of imported soil material, particularly that which will be used to fill planters and planting beds. The availability of fertilizer components, minerals, and trace elements to the plant is directly controlled by the pH factor. In general, a pH range between 6.5 and 7.5 will allow most plant nutrients to be chemically available to ornamental plants. In other words, the soil chemistry is controlled to a great extent by the soil's pH.

Fertility

A fertility test may be referred to as an *agricultural suitability* test. The test assesses a soil's nutrient availability in terms of the major components—nitrogen (N), phosphoric acid (P_2O_5), and potash (K_2O). These three nutrients are generally indicated as N, P, and K, and their chemical composition is more complex than can be described here. Several other nutrients, minerals, and metals will be assessed by the test; for example, sulfur, magnesium, calcium, manganese, iron, boron, zinc, copper, and molybdenum Most of these constituents are referred to as *trace elements*, a term descriptive of their quantity in the soil and not of their importance to plant growth.

Moisture

Moisture and soil texture are inseparable as they relate to plant growth. Selection of plant species, installation requirements, costs, installation timing, equipment used, maintenance, conditions, the guarantee of plants, and the length of an installation period are directly related to the soil's reaction to moisture. An irrigation system designer must also have knowledge of the soil's moisture capacity before design can proceed. Irrigation design will be based on a soil's ability to infiltrate, hold, and lose water in relationship to the plant's transpiration characteristics and the time available for system operation.

Information needs often extend to laboratory tests and interpretations of a soil's wilting point, field capacity, and available moisture at various depths of the soil profile.

Testing

Most nutrient testing of soils will be chemical in nature and most suitable for laboratory analysis. In all instances, the interpretation of data must be made by a person knowledgeable about local soil conditions and the nature of proposed plantings. Perhaps the best information can be gathered by combining laboratory and greenhouse testing. After chemical analysis indicates soil properties, a series of variable nutrient tests are run on indicator plants. Following assessments of a plant's appearance and growth habits, and perhaps tissue tests, a very firm recommendation of nutrient requirements can be made by a consultant. Tests of live plants as indicators are expensive and time consuming but, in the long run, can save thousands of dollars worth of fertilizers and minerals that often are specified and applied as much from habit as from knowledge.

When a site is to undergo extensive grading with mass fills and cuts, it can be time saving and informational if soil borings taken for soil engineering tests are also used for agricultural suitability sample testing. A designer will need to know the horticultural properties available in the B and C horizons, as well as the A horizon.

3.12.2 Engineering Properties

A soil's engineering properties will generally be revealed through a series of tests that serve as indicators. Tests are necessary to predict potential physical soil strength in compression and shear, consolidation, moisture content, dry density, plasticity, percolation, compaction, and expansive character, and chemical tests relating to soil–water reactions are also required.

In general, testing can be accomplished by rough approximations conducted and observed on the site or through relatively complex laboratory studies using special test equipment. In either case, the nature of soils is quite variable and any test will be subject to interpretation.

When so authorized, a designer will contact or recommend a soils engineer to serve as a consultant. The soils engineer conducts various tests and prepares an initial report regarding the characteristics of the soil(s) within a site and makes recommendations as to alternative foundation design, and perhaps grading or drainage procedures. The initial report may be supplemented by other requested reports and recommendations prepared during the site work. Many engineering firms also continue, during the work's progress, to provide field observations, surveys, and testing of a contractor's work for compliance with technical specifications.

When a part of the site work includes the construction of load-bearing structures, the soil investigation will include some type of vertical exploratory excavation. Large-diameter pits may be dug for personal observation of strata or the drilling of cores may provide a profile of soil types. In either event, excavated soil samples are made available for testing. During any excavation, the various soils are recorded as to their depths, water encountered, and types. A soil engineer will portray the profile in a graphic form called a *soil log*. The soil log and the various soils from the excavation will then be interpreted. It is this information that is part of an owner's information released to a contractor(s) prior to bidding on any earthwork.

It is necessary that a soil engineer know approximately the dimensions, locations, and character of the designed structures so that test borings may be located and mapped. The number of borings (or even the need for borings) is a professional opinion of an engineer and relates very much to prior knowledge of the site and the nature of any structures.

Although many of the necessary tests are identified and specified as to procedure by the American Society for Testing and Materials (ASTM) or the American Association of State Highway Officials (AASHO) or are otherwise standardized, it is best to allow the engineer freedom to choose the number of types of tests necessary for design and construction decisions. On occasion, the number and types of tests will be dictated by the owner, usually a public entity.

In general, the development of a scope of soil engineering work will be similar to that of any other professional involved with the site. It is, however, always a good idea to frame an agreement for services between the engineer and, usually, the owner. An agreement might include a scope of services that delineates the owner's needs—for example, the provision of a report, interim reports, and final report (postconstruction), and the contents of the report as a precise soils map, soil logs, interpretation of soil characteristics, boring map, methods of excavation, test run and procedures used, interpretations, opinions, technical specification language, concerns, foundation design, bedrock geology, organic matter, agricultural suitability, site conditions, researched information, settlement estimates, and soil classification system used. In addition, an engineer's possible involvement in observing the site work's progress, meetings, and measurement duties should be clarified in relation to the type of fee devised.

In many respects, the owner's decision regarding the quantity and quality of information will relate to the manner in which a contract agreement will be arranged and the owner's inclination to provide information. A lump sum construction contract, for example, may force a high contingency and risk upon a contractor if little information is made available. However, a minimal amount of information is often necessary to determine the soil's structural bearing capacity for architectural design purposes.

3.13 LEGAL RESTRICTIONS

An owner's right in a particular site might be viewed as a "bundle" of rights, that is, a variety of rights commonly called ownership. Generally such rights in ownership are conveyed and restricted by the states, and occasionally by the federal government. Zoning ordinances, for example, are developed by a city on the basis of enabling legislation by a state. Public zoning, either city or county empowered, will be one piece of information that may affect the site plan and its implementation.

The owner must provide, or authorize a designer

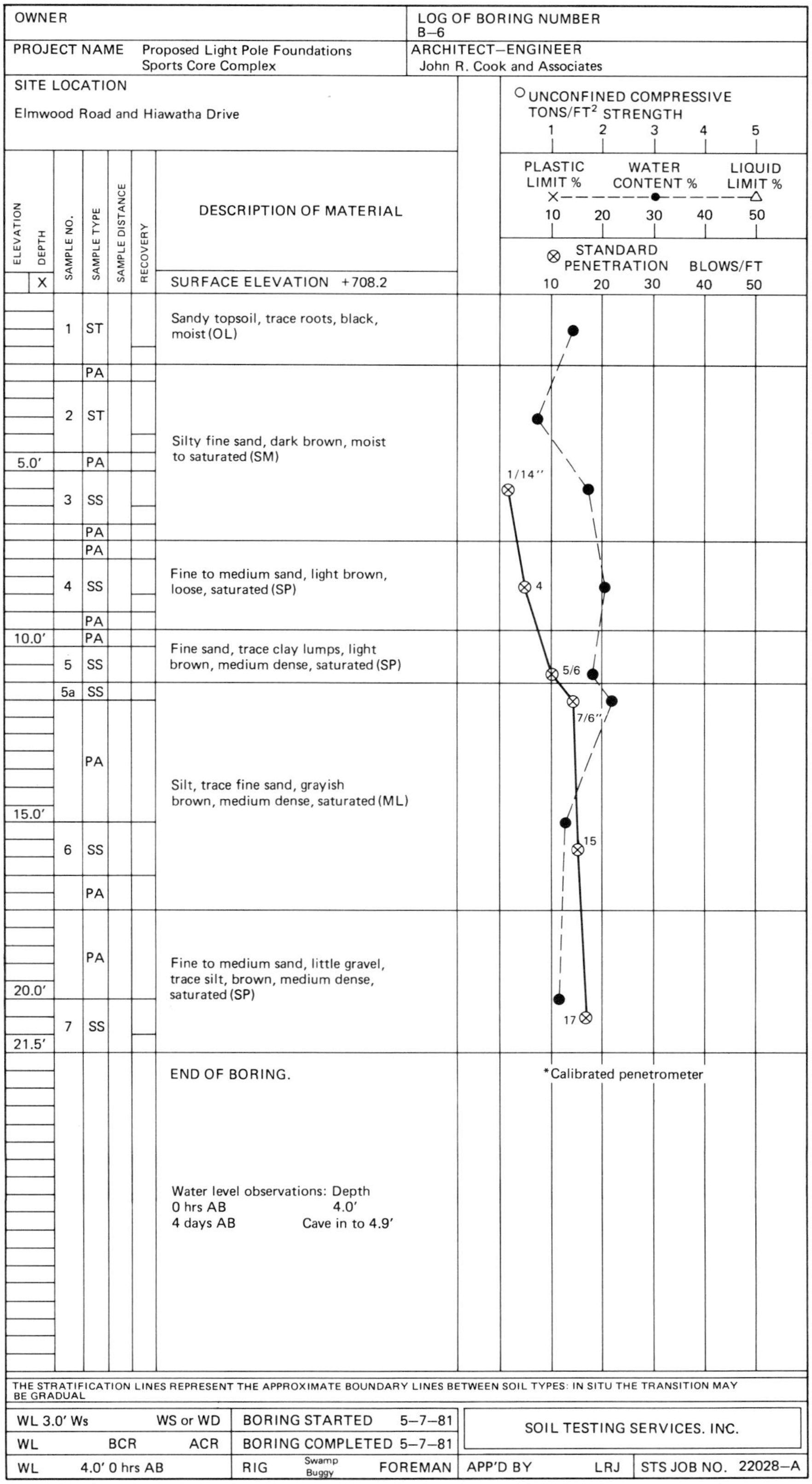

OWNER						LOG OF BORING NUMBER
						B–6

| PROJECT NAME | Proposed Light Pole Foundations Sports Core Complex | | | | | ARCHITECT–ENGINEER John R. Cook and Associates |

SITE LOCATION

Elmwood Road and Hiawatha Drive

○ UNCONFINED COMPRESSIVE TONS/FT² STRENGTH

```
1     2     3     4     5
```

PLASTIC LIMIT % ✕— — — WATER CONTENT % ● — — — LIQUID LIMIT % △

```
10    20    30    40    50
```

⊗ STANDARD PENETRATION   BLOWS/FT

```
10    20    30    40    50
```

ELEVATION DEPTH	SAMPLE NO.	SAMPLE TYPE	SAMPLE DISTANCE	RECOVERY	DESCRIPTION OF MATERIAL
X					SURFACE ELEVATION +708.2
	1	ST			Sandy topsoil, trace roots, black, moist (OL)
		PA			
	2	ST			Silty fine sand, dark brown, moist to saturated (SM)
5.0'		PA			
	3	SS			1/14"
		PA			
		PA			
	4	SS			Fine to medium sand, light brown, loose, saturated (SP) 4
		PA			
10.0'		PA			
	5	SS			Fine sand, trace clay lumps, light brown, medium dense, saturated (SP) 5/6
	5a	SS			7/6"
		PA			Silt, trace fine sand, grayish brown, medium dense, saturated (ML)
15.0'					
	6	SS			15
		PA			
		PA			Fine to medium sand, little gravel, trace silt, brown, medium dense, saturated (SP)
20.0'					
	7	SS			17
21.5'					
					END OF BORING. *Calibrated penetrometer

Water level observations: Depth
0 hrs AB   4.0'
4 days AB   Cave in to 4.9'

THE STRATIFICATION LINES REPRESENT THE APPROXIMATE BOUNDARY LINES BETWEEN SOIL TYPES: IN SITU THE TRANSITION MAY BE GRADUAL

WL 3.0' Ws		WS or WD	BORING STARTED 5–7–81	SOIL TESTING SERVICES. INC.	
WL	BCR	ACR	BORING COMPLETED 5–7–81		
WL	4.0' 0 hrs AB		RIG Swamp Buggy FOREMAN	APP'D BY LRJ	STS JOB NO. 22028–A

BL:I

Figure 3.4 Typical information as to structural properties of a soil. This "log of borings" is one of several studies undertaken by a soils engineer to determine the ultimate size of footings, post and column bases, piers, and other structures that depend upon support from soil. (Courtesy John R. Cook Associates, Inc., landscape architects, Rockford, Illinois)

to provide, information respecting zoning. Some owners of large developments may wish personally, and with legal counsel, to handle all zoning matters. However, the ordinary site will simply lie within a certain published zone that must be legally recognized in the design and construction phase.

Deeds to land sometimes carry restrictive private and public land use controls called covenants. The language and purpose of such covenants are to restrict or protect a specific site from a specific use, that is, some of the bundle of rights are removed or restricted. If such covenants "run with the land" and are enforceable by law, a designer must recognize them in site planning and detailed implementation. A typical planned unit development, for example, is created by local zoning ordinance but may also carry restrictive or protective covenants. One existing industrial park within a planned unit development carries protective covenants, for instance, that control setbacks, completion and finish of site work, excavations, planting species, planting locations, planting completion, maintenance, signs, parking area design, loading and storage of materials, site plan approvals, and permitted and nonpermitted land uses. In most instances, a covenant will apply to an entire site, boundary line to boundary line.

Easements are public or privately held rights acquired by a person who does not own the land upon which an easement exists. In other words, the owner of a site willingly gives up one of the rights ordinarily granted by ownership. Quite often, an easement has dimension—for instance, a strip of space 10 feet wide granted by a homeowners' association to a utility company for pipe burial. Such an easement might also grant the utility company a right of access over, and only over, the width and length of the easement.

The existence of an easement is usually found by a licensed surveyor during the course of property survey and title search. The easement will be plotted on the site's boundary line survey. Each easement will relate to specific usage of the easement. A designer must know precisely what the easement allows or restricts. For example, many easements grant a right of access that, essentially prevents the construction or installation of anything within the easement that will restrict accessibility.

Relatively recent national energy source problems have led to a renewed interest in developing easement rights for light, air, and views. These types of easements must be carefully researched as they will be quite new in each community or county. A designer may find that a specific site is restricted as to how high buildings may be constructed or plants may grow.

SELECTED READINGS AND REFERENCES

Brown, Curtis M. *Boundary Control and Legal Principals.* New York: Wiley, 1969.

Christiansen, Monty L. *Park Planning Handbook.* New York: Wiley, 1977.

Fletcher, Gordon A., and Vernon A. Smoots *Construction Guide for Soils and Foundations.* New York: Wiley, 1974.

Kissam, Philip *Surveying Practice.* New York: McGraw-Hill, 1971.

Lynch, Kevin *Site Planning,* 2nd ed. Cambridge, Massachusetts: M.I.T. Press, 1971.

Martin, Alexander C., et al. *American Wildlife and Plants: A Guide to Wildlife Food Habits.* New York: Dover, 1951.

Marsh, William M. *Environmental Analysis for Land Use and Site Planning.* New York: McGraw-Hill, 1978.

Munson, Albe E. *Construction Design for Landscape Architects,* New York: McGraw-Hill, 1974.

Rubenstein, Harvey *A Guide to Site and Environmental Planning,* 2nd ed. New York: Wiley, 1980.

Schroeder, W.L. *Soils in Construction,* 2nd ed. New York: Wiley, 1980

Chapter Four
Construction Plan

*Any contractor walking around the job site with a smile on his face
is subject to review of his bid.*

Anonymous

4.1   THE DRAWING AS INFORMATION

A construction site plan contains information that
pertains to the construction of such features as
paving, benches, fences, and walls. Ordinarily this
drawing will be a single-purpose sheet identifying
the types of construction materials in general terms,
an indexing system relating to construction details
in the set of drawings, and an overall delineation of
the site construction features. On occasion, the
sheet will also delineate vertical or horizontal
dimensions, construction notes, and miscellaneous
information. The drawing also relates various construction to technical specifications, serves as a
basis for estimating construction costs, and provides a count of individual features to be constructed or installed.

If an overlay drafting system is used, the construction plan is used as base information upon
which the grading, drainage, plantings, utilities, and
lighting plans will be prepared.

4.2   ACCURACY

The drawing's accuracy can be no greater than that
of the drafter's skill and the original site survey.
Because the drawing is often used to estimate areas
and linear measurements, it must be as accurate as
possible. Both the drafter and contractor should be
concerned about the possible graphic inaccuracy
that various plan scales may entail. For example, at a
scale of $\frac{1}{8}$ inch = 1 foot, a potential minimum scale
human error of 3 inches exists; at a scale of 1 inch =

20 feet, a possible scaled human error of 1 foot exists;
and at a scale of 1 inch = 30 feet, the possibility of a
$1\frac{1}{2}$ % = foot scaled error exists. When possible,
dimensional information should be taken from the
horizontal control plan and substituted for scaled
dimensions.

4.3   DRAWINGS AND CONSTRUCTION
COORDINATION

A construction drawing for site development will
often span the gamut from very-small-scale site
plans to large-scale details. The designer, drafter,
and contractor must be aware that a series of
constantly increasing scales of drawings may be
necessary to delineate precise construction quality
and quantities. Figures 4.1–4.4 illustrate such a
progression of scale that might be used. Figure 4.1 is
a portion of a small-scale site plan showing some
construction elements as a group. Figure 4.2 is the
same plan view, but with an enlarged scale and a
more specific portion of the construction delineated. Figure 4.3 is an elevation of the plan view of
Figure 4.2 of sufficient graphic size for dimensions to
be added. Figure 4.4 is a greatly enlarged view of a
connection for a portion of the construction. There
may be several details such as Figure 4.4 necessary
to describe the construction of various parts of the
structure.

Note that the small scale of Figure 4.1 is not
suitable for much specific information. Its purpose
is to guide the reader to sheets of the documents
where enlarged views with more specific informa-

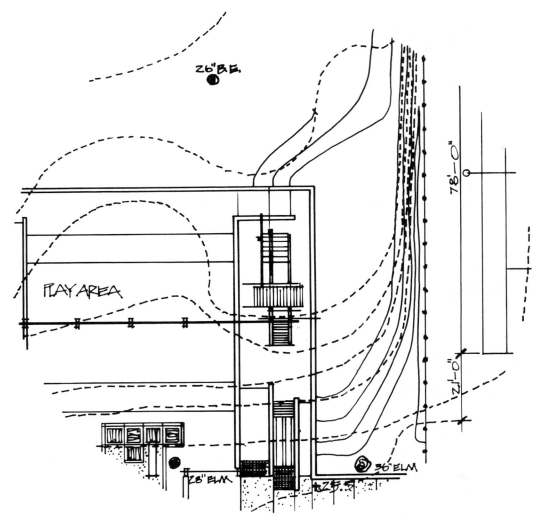

Figure 4.1   Portion of a small-scale site plan.

tion may be seen. Such a site plan also serves to maintain a perspective of exactly where the various items are to be constructed.

Both Figure 4.2 and Figure 4.3 might be referred to as details because they are enlarged portions of the site plan. Each of the drawings is graphically large enough to contain specific and detailed information regarding construction materials, dimensions, and sizes. Note that these scales are too small effectively to delineate connections among materials but are large enough to indicate material and member arrangements.

Figure 4.4 is an example of an enlargement of a specific portion of the structure. At this point, there is sufficient room for the addition of material tex-

tures, construction notations, and structural connection methods to be shown.

In all of the above examples, the choice of necessary scales and views is a matter of judgment. In most instances, it comes down to deciding just how complicated the construction problems are and whether or not several graduations in scale drawings will help the contractor or simply will add complexity to the drawings. Sometimes it may be best to jump graphically from the small-scale site plan directly to an enlarged detail similar to Figure 4.4, if the detail can be obviously located on and oriented to the site plan.

Several items should be noted with regard to these example details. Figures 4.2 and 4.3 both contain

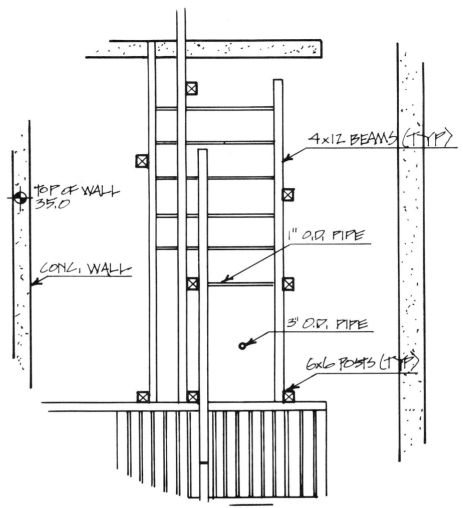

4×12 BEAMS (TYP)

TOP OF WALL
35.0

1" O.D. PIPE

CONC. WALL

3" O.D. PIPE

6×6 POSTS (TYP)

Figure 4.2 Portion of a site plan at a scale larger than that of Figure 4.1.

specific notations as to the sizes of wood members, but the detail, Figure 4.4, does not size these members again. If both scales of details contained information as to sizes or other information, it would be redundant. Which of the details carries the information is not as important as the avoidance of the wasted time and confusion that may result from drafting redundant information on two different scales of drawings. For example, Figures 4.2 and 4.3 identify a member as a 6 × 6 post, and detail Figure 4.4 simply identifies the same member as a post. It is extremely important that the same term—that is, post—be used by the drafter to identify every similar member throughout the drawing set. Figures 4.2 and 4.3 also identify the 6 × 6 post as TYP. (typical), the abbreviation meaning that every post in the struc-

ture is a 6 × 6 dimension. This is a proper procedure to save time and redundancy but the term must be used *only* when no other post differs in dimension.

Terminology used on drawings also relates to other portions of the contract documents. For example, if the term *post* is used on the detail drawings, the same term must be used in schedules, legends, and specifications. The size of a post may change in various details but the term must be used only to identify one specific purpose and installation. The contract documents become quite confusing if a drafter uses "post," "column," and "support" to identify the same kind of structural member. The choice of which term to use is not as important as is the consistent use of that term throughout all the contract documents. Although the need to reduce

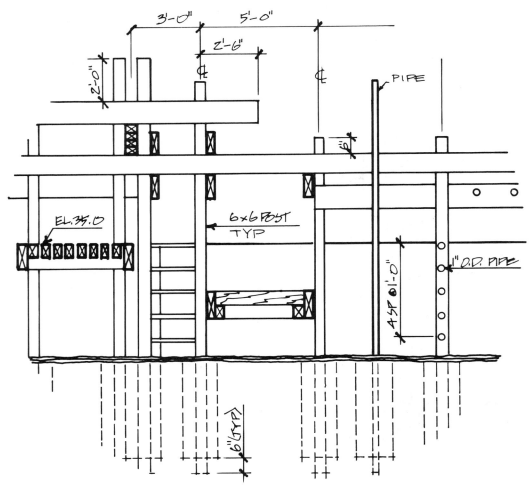

Figure 4.3   Elevation view of the site plan shown in Figure 4.2.

redundancy and to avoid conflicting terminology may seem obvious, both negative procedures commonly creep into contract documents.

4.4   INDEXING AND REFERENCING

Drawings are essentially a book of drawings. Each of the sheets delineates some aspect of the site work and they open in a manner similar to the pages of a book. Because of the booklike format, a designer must often refer the reader to information contained in the various pages. Each set of drawings will have its own complexity determined by the number of sheets and the number of different subjects delineated. Major site development projects may contain several sheets of site drawings, architectural drawings, utility drawings, horizontal and vertical

controls, and detail drawings. A contractor and subcontractor must be able to find their way through such information quickly, during both the bidding period and construction and installation. The process of locating information in the various sheets is commonly called indexing or referencing.

4.4.1   Basic System

Each detail is identified by a single letter or number. The site plan identifies where each of the details will be constructed or installed. Each of these graphic symbols indicates meaning by the manner in which it is drawn. Figures 4.5−4.8 illustrate typical graphic symbols and their ordinary meanings. Note that the indexing system interrelates plan, section, and elevation views among themselves as well as with enlarged detail views. Each view is identified by a

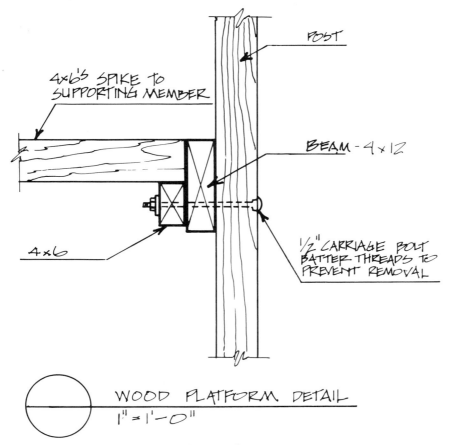

Figure 4.4   A detail of structural connection used in Figure 4.3

title beneath the detail. Each symbol letter (or number) appearing on a view references to a specific detail title.

4.4.2   Common System

A second system is necessary whenever the number of drawings and details becomes so complex that the various views must be scattered among the various sheets. Figures 4.9 through 4.12 illustrate one type of such a system. Note that the symbol identifying where the particular detail may be located matches both the sheet number and the title number of the particular detail appearing on that sheet. For example, Figure 4.9 indicates that a sectional detail of the object may be located as detail 2 on sheet 4. The top number in the circle represents the detail title number and the lower number is the sheet number. The symbol 2/2 then references to a sectional detail 2 on sheet 2. Of course, each of the

sheets would carry many more views and details than shown in these figures. A legend should appear on the drawings so that the contractor understands which of the numbers (or letters) represents title and sheet numbers.

4.4.3   Cross-Indexing

A cross-indexing system is very similar to the common system, except that the manner of identifying the various views is a bit more complex. The common system does not allow a reader to relate a detail to the site plan or an enlarged view. In other words, a shortcoming of the common system is that it progresses from the small scale to the large scale with no way of reversing the indexing flow. A cross-indexing system will allow a reader, for instance, to relate a detail to a construction plan so that the progress moves from large-scale to small-scale drawings.

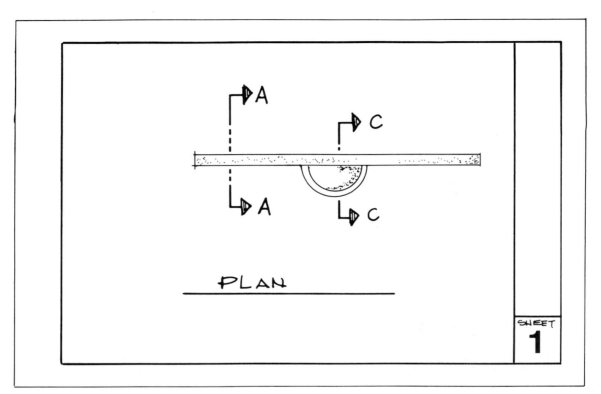

Figure 4.5   Example of a plan view with a simple reference system.

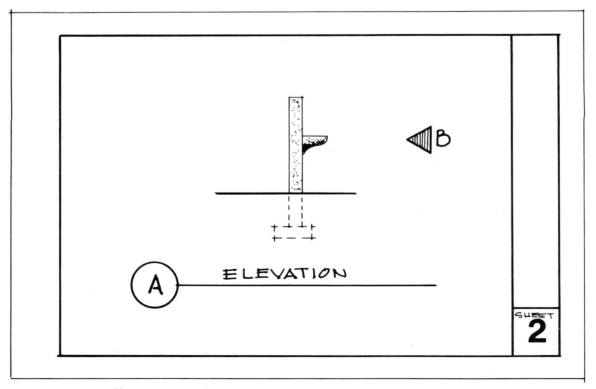

Figure 4.6   Example of an elevation view with a simple reference system.

60

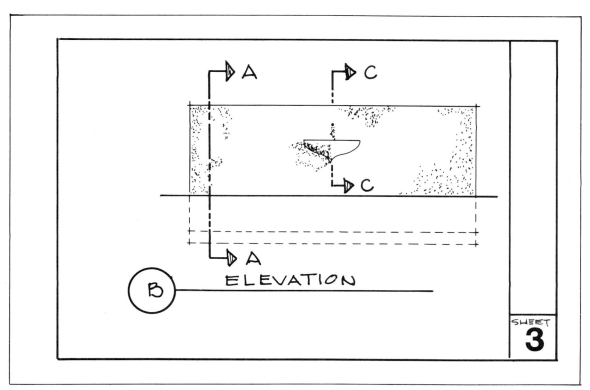

Figure 4.7   Example of an elevation view with a simple reference system.

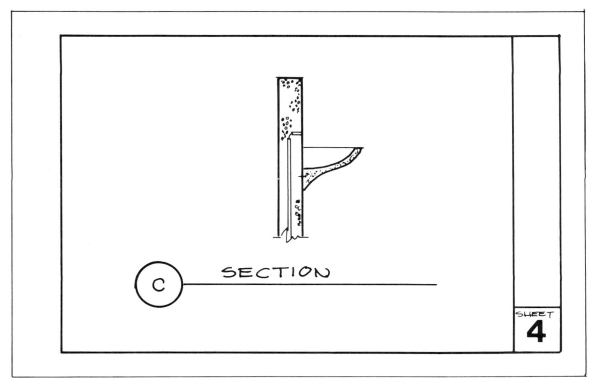

Figure 4.8   Example of a sectional detail keyed to Figures 4.5 and 4.7.

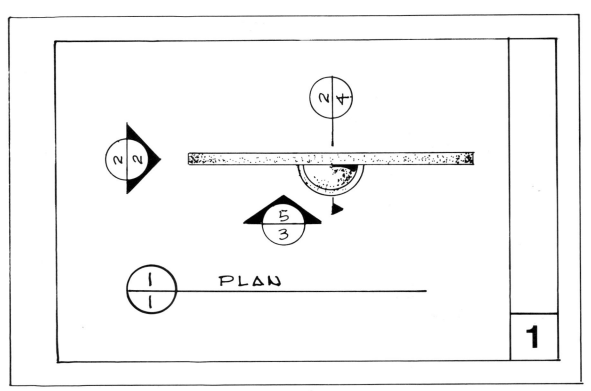

Figure 4.9 Example of a plan view reference to Figures 4.10, 4.11, and 4.12.

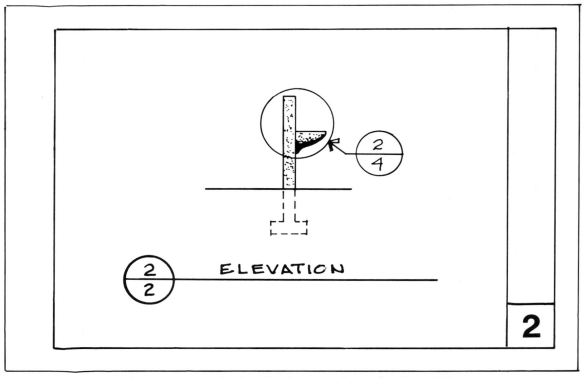

Figure 4.10 Example of an elevation view referenced to Figure 4.12.

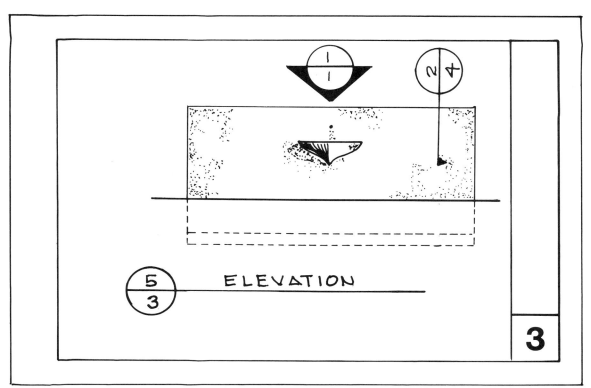

Figure 4.11   Example of an elevation view referenced to Figures 4.9 and 4.12.

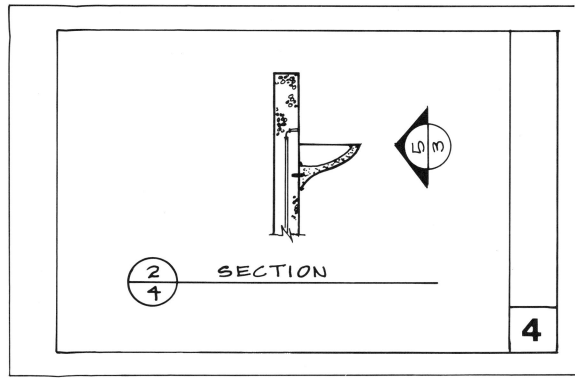

Figure 4.12   Example of a sectional detail referenced to Figure 4.11.

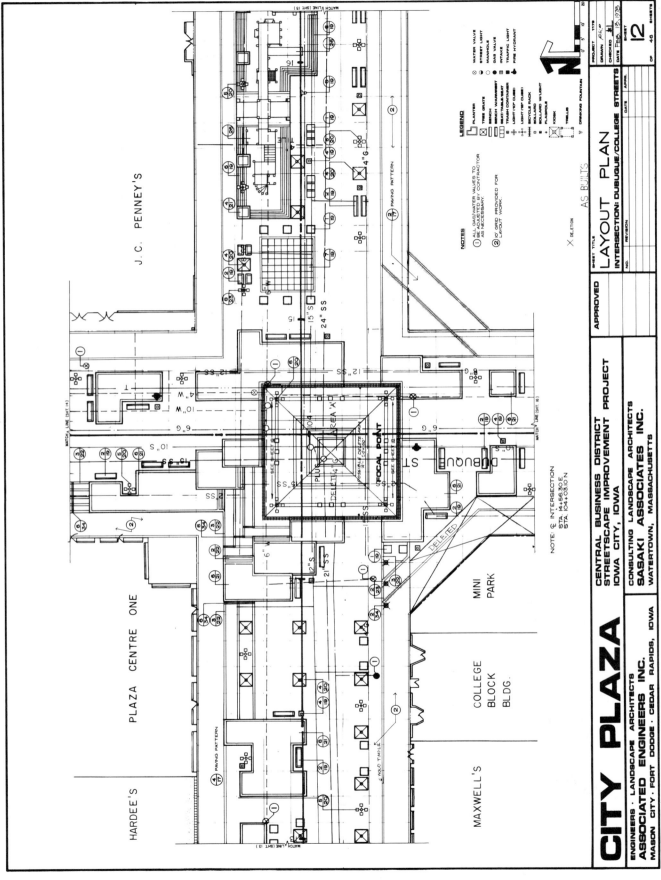

Figure 4.13   Example of a reference system in plan view relating construction to details among various other sheets. (Courtesy Jack Leaman, landscape architect, Ames, Iowa)

Figures 4.14–4.17 illustrate a cross-indexing system in a stylized portion of a set of drawings. Sheets L-1 and L-2 (Figures 4.14 and 4.15) are plan views of a building site with small details of construction shown on sheets L-8 and L-9 (Figures 4.16 and 4.17). Note that detail 3/L-8 is found on sheet L-8 and detail 3 on sheet L-8 is identified as found on sheet L-1. Note also that detail 1 on sheet L-9 refers to sheet L-8 for additional information on detail 1/L-1.

Although such a cross-indexing system may appear complex, the method is quite necessary to the contractor and the many drafters who may be using a large set of drawings with many sheets. A subcontractor, for example, may be responsible for fabricating one detail of the construction and may need to know where the detail is used, how many units must be fabricated, and the surrounding context of the detail. Without a cross-indexing system, it would take time to search through all of the site, architectural, or engineering drawings to find how the detail fits into the whole. It would be a bit like looking through a microscope to try to figure out where a small view fits into the whole scene.

4.4.4 Adaptations

Whenever site work drawings are a part of architectural, utility, electrical, or engineering contract drawings, it is a convenience for everyone involved if each of the various subject drawings is identified. For example, the site work may be identified by the letter "L." Each sheet that contains site information thus can be quickly identified and separated from other types of construction or installation work (see Figures 4.14–4.17).

Common letter designations for various subjects are "L" for landscape, "A" for architecture, "S" for structural, "I" for irrigation, "PL" for planting, "E" for electrical, "EL" for exterior lighting, "P" for plumbing, "G" for grading, "SD" for storm drain, and "D" for dimensioning. When such categorization is used, the letters are prefixed to each sheet number, for instance, L-1 or L-2, and used in each of the indexing symbols and detail titles.

4.5 CONNECTIONS

Contract documents have been defined as the drawings and specifications taken together. By this definition, it becomes extremely important that clear connections exist among the various drawings and the written specifications. Unfortunately many mistakes exist in contract documents as a result of the failure to clarify connections between drawings and specifications.

4.5.1 Sources of Problems

Problems arise from various sources:

1. *Use of different terms for the same item.* An example of this is the use of the terms *paving* on drawings and *concrete* in the specifications. The term *concrete paving* is used on a drawing if several types of paving materials exist. The specification heading *concrete paving* will then complement the drawing.

2. *Insufficient use of graphic index symbols.* Contractors are often forced to guess at the intention of the designer with respect to where a detail and material are to be used. Designers may assume a contractor to be a mind reader.

3. *Redundancy of information among drawings and specifications as potential conflict.* Among examples of such redundancy are dimensioning a detail and repeating the information in the specifications; specifying the manufacturer's name and model numbers and repeating the same information on the detail drawing; sizing the same wood member on a site plan, a detail, and a schedule; drawing details of the same area at different scales; and using construction notes that repeat the same directions given in specifications. Other examples include redrawing a manufactured item with instructions while the specifications reference to manufacturer details and directions; repeating dimensions on a horizontal control plan and an enlarged detail; repeating sizes of items throughout the various views and scales of the drawings; and drafting both sides of all symmetrical items.

4. *Drafting and writing specific directives that conflict with the conditions of the contract.* An example is a drawing notation that directs the contractor to perform a specific procedure, whereas the specification language dictates material performance and the general conditions delegate all methods and procedures to the contractor.

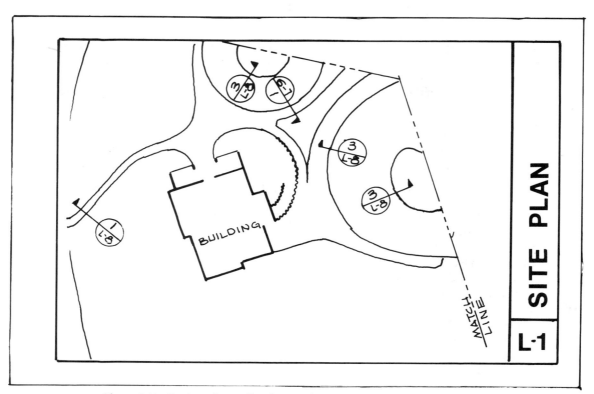

Figure 4.14　Portion of a small-scale site plan appearing on a drawing sheet L-1.

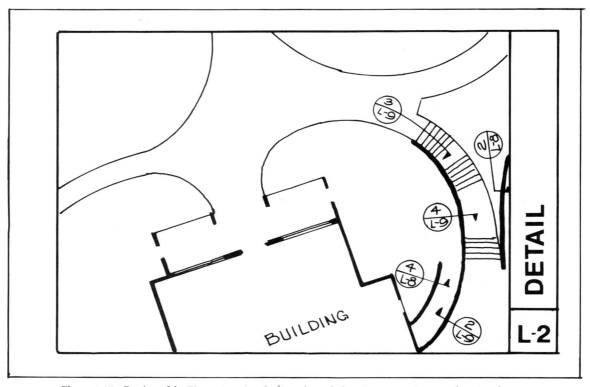

Figure 4.15　Portion of the Figure 4.14 site plan in enlarged plan view appearing on a drawing sheet L-2.

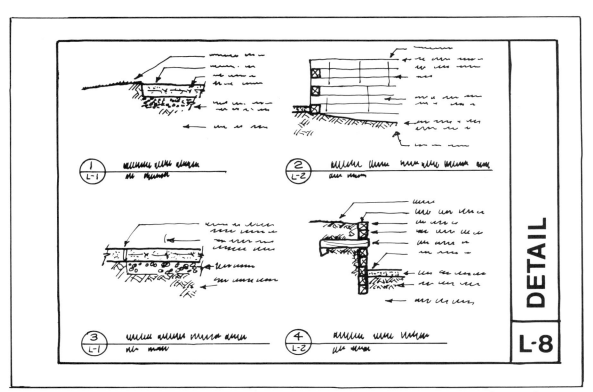

Figure 4.16 Typical sheet upon which details are drawn and numbered. Note that the reference of some items appearing on Figures 4.14 and 4.15 are located on this sheet L-8.

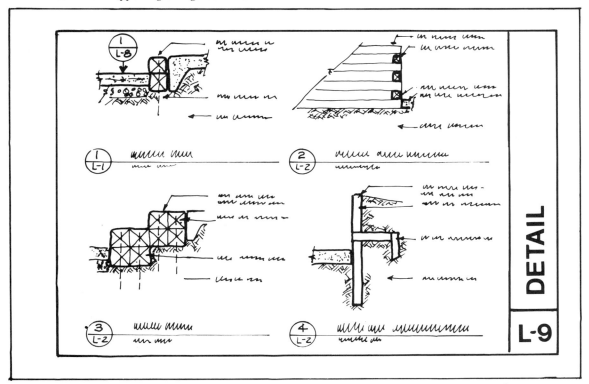

Figure 4.17 Typical sheet upon which details are drawn and numbered. Note that references of some items appearing on Figures 4.14 and 4.15 are located on this sheet L-9, and that detail 1/L-1 also references from sheet L-9 to sheet L-8.

5. *Overdrafting of information in such a graphically "artistic" form that construction information* may be obscured. Communication with an owner in the form of presentation drawings must be distinct from the form in which information is conveyed to a contractor. Simplicity and clarity of information are key features of graphic communication with a contractor. The graphic use of shade, shadow, dark textures, extraneous line work, and other forms of artwork consumes drafting time and decreases the legibility of drawings.

6. *Use of manufacturers' names on drawings instead of in specifications.* Substitutions of manufactured items actually incorporated in the work produce a detail with useless and conflicting information.

4.6 WOOD STRUCTURES

Lightweight engineering and architectural structures may be part of the site development. Whether or not the contract documents reflect a dominance of site work or structures depends upon the relative magnitude of the site work and architectural construction. In any event, the contract documents should follow a format similar to that of architectural drafting and arrangement with only the number of sheets and the specification content reflecting the dominant character of construction or site installation work.

Wood structures are a common feature of many site developments. Such structures include lightweight decks, shading structures, and small frame buildings. Graphic formats closely follow common drafting conventions of architecture, however, the small size and open character of many structures may call for slightly revised approaches to their graphic reproduction.

4.6.1 Supports

Many lightweight and open structures are placed on piers or posts set vertically in holes excavated through the finish grade. Standard architectural conventions, such as the plan, sectional, and elevation views, are useful in the delineation of these structures.

A plan view should be based on the support post installed in the center of a structure's module. All dimensions will usually be center to center of each post position. If a post penetrates the finish grade

and is set in concrete, a concrete collar should be indicated as a hidden line. If a post is set on a pier, the pier itself should be shown as a solid line along with the post. A drafter must keep in mind that this particular plan view will be a horizontal section; that is, vertical members that protrude above the finish grade are boldly drawn and identified with an "X" as a symbol for a wood in cross section.

The following must be checked for potential conflict.

1. Can the posts be dimensionally located from an existing structure, property line, or some other fixed starting point?

2. Are the elevations of piers coordinated with the grading plan elevations and structural elevations, or will the structural plan view serve as vertical control of the pier elevations?

3. Do elevations or details of the structure coordinate with the finish grade, pier size and location, pier depth into soil, and pier height above the finish grade?

4. Are there any dimensions on cross sections or elevations that may conflict with the length of posts relative to the structure's finish floor elevation and the pier or finish grade elevations?

5. Does the grading plan reflect the finish grade around piers or posts, the area beneath the structure, storm drainage, the structure's finish floor elevation—that is, which drawing controls these pieces of information?

6. Do details control the character of the connection between pier and post?

7. Do specifications control the compaction of earth?

8. Does the design coordinate with codes relative to wood construction above and below the finish grade?

9. Are there any redundant or conflicting dimensions between the enlarged plan detail, sectional details, and the overall horizontal control plan?

10. Are the locations and vertical elevations of any steps coordinated with the plan view, finish floor elevation, and details of the step connection to the finish grade?

11. Are plantings and irrigation coordinated with the area beneath the structure?

12. Are there any subsurface utilities existing or proposed which support excavations might penetrate or with which they might interfere?

13. Do the contract's conditions cover subsurface excavations and responsibilities for subsurface characteristics?

14. Is there any conflict between the location and depth of structural supports and subsurface footings of existing or proposed retaining or other walls?

4.6.2 Structures

A structure will ordinarily be shown as plan, section, and elevations. Each of these views must have a precise purpose in order to avoid conflicting information.

A plan view can often be drawn as one view with several layers exposed to viewing from above. For example, one layer might be decking, a second joists, and a third might expose beams and posts to view. If the structure's plan view is symmetrical, it may be possible to save time by drafting only half of the structure. If this is done, the drafter must be sure to delineate the entire perimeter of the floor plan and to identify the centerline of the structure. A note should be included on the drawing that the view is one half of the structure and that it is to be constructed "symmetrically about centerline." The following should be checked regarding drawings and specifications, in addition to the previous check list.

1. Which document controls the size, wood type, and finish of the decking or other surface? Is there any conflicting information or does one document or schedule fix each bit of information?

2. Are all of the structural members sized and identified on one drawing or may there be potential conflict and redundancy among several sources?

3. Do the views, member positions, and terminology used on details match those shown on the various plan and elevation views?

4. Is there any information on an elevation view—for instance, material—which conflicts with a plan or detail?

5. Does any elevation view show vertical dimensions that may conflict with spot elevations shown on the grading plan?

6. Do any enlarged plan view dimensions conflict with the site's horizontal control drawing?

7. If the structure stands free of any existing structure, is there dimensional information on the site's horizontal control drawing that locates the structure?

8. Have utilities been shown on the site plan as

entering the structure? Is there an enlarged diagrammatic drawing of the interior utilities.

9. If wood material is to be artificially impregnated with wood preservatives, has the chemical preservative been checked for its potential toxic effect on plant materials?

10. Are all structural connectors specified as to type, material, and installation?

11. Do drawings or schedules and/or codes control the sizes of structural connecting devices?

12. Are all necessary index symbols located and index figures completed for the graphic interrelation of the site, construction plan, enlarged views, graphic elevations, and details?

13. If there are multiple prime contracts, are responsibilities for setting, boring, or other work necessary to utilities or lighting defined?

4.6.3 Construction Specifications

The following information must be a part of the specifications.

1. *Species of wood.* Most often, this will refer to the common name of a particular tree, however, a few wood associations have developed species groups. For example, the Western Wood Products Association groups Douglas fir and larch as a single entity on the basis of appearance and strength. In many cases, lumber yard personnel cannot visually distinguish among groups.

2. *Grade.* Commercial lumber grades are generally defined as that quality of material that can be supplied to the project by a specific wood association. Two ways of grading are used by the wood associations—visual and structural. If a designer needs to control the esthetic aspect of a structure, the grade specified should be one that identifies visual quality with normal strength. For example, the California Redwood Association's grade Construction-Heart is visually graded for minimal knots and imperfections, whereas the Western Wood Products Association's grade #2 is also graded as to limitations of knot diameter, spacing, condition, and imperfections.

Structural grading differs from visual only in that physical tests determine the average allowable stress within the piece of wood. The wood's appearance may actually be better than that of nonstructural grading material. For example, the Western Wood Products Association grades "select struc-

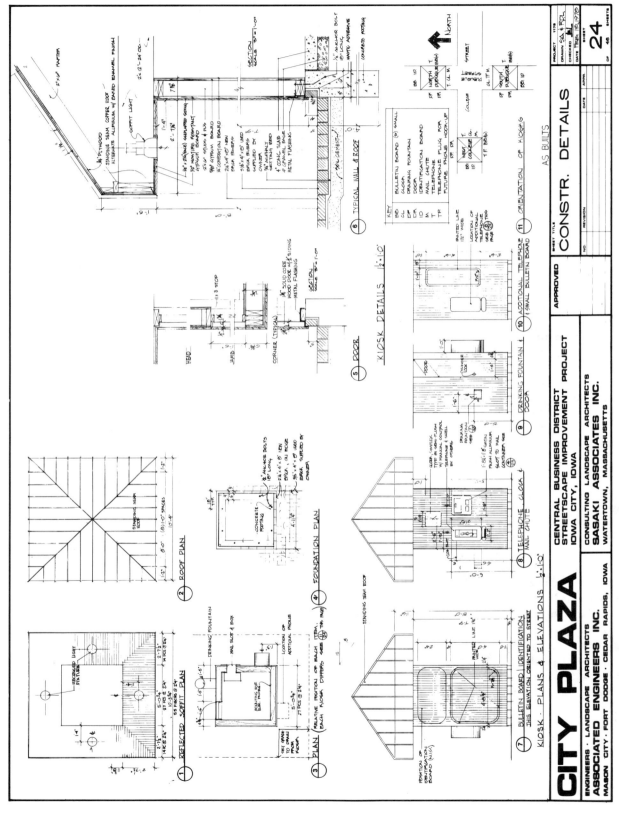

Figure 4.18 Typical construction detail sheet with each detail numbered. Note the orderly modular arrangement of the details and the sequential numbering of details from left to right and top to bottom of the sheet. (Courtesy Jack Leaman, landscape architect, Ames, Iowa)

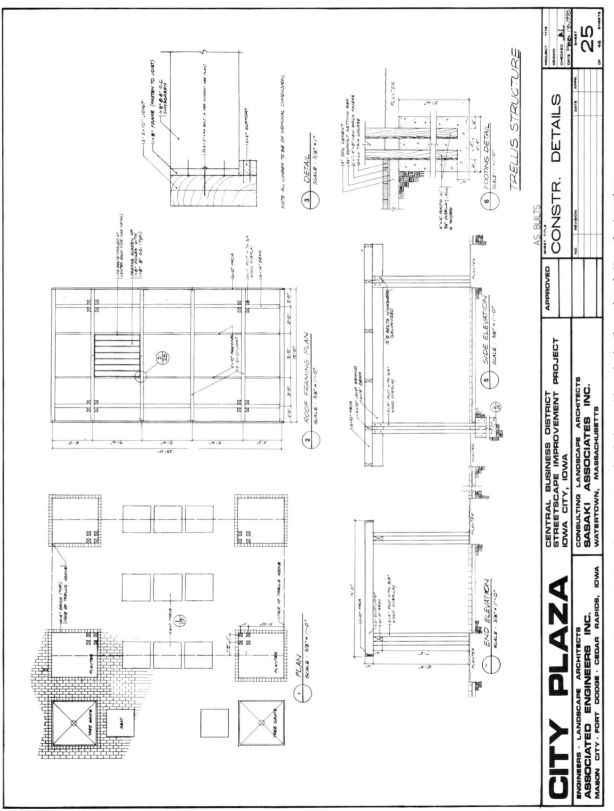

Figure 4.19 Typical construction detail sheet with each detail numbered. Note the intrasheet reference to enlarged detail as well as that reference from detail 1 to Figure 4.20 for a "seat table" detail is found as 1/19. (Courtesy Jack Leaman, landscape architect, Ames, Iowa)

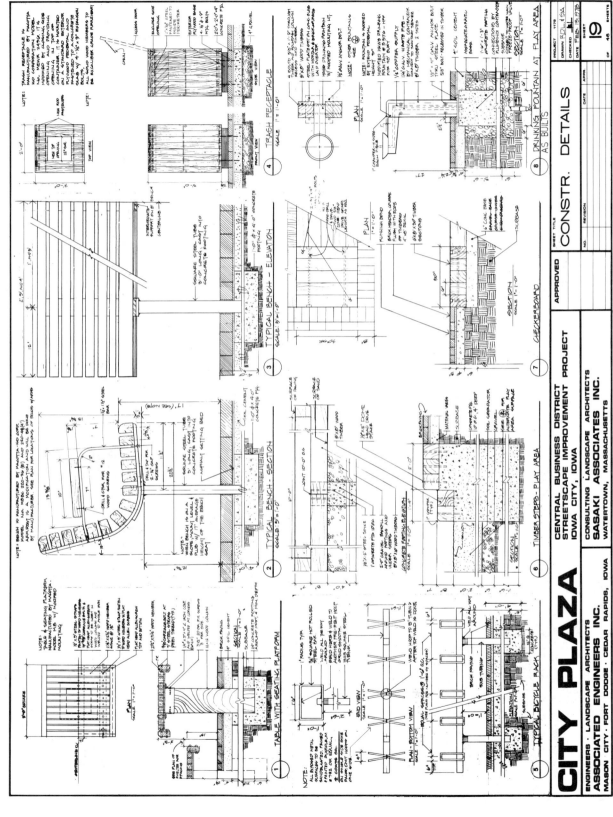

Figure 4.20 Typical construction detail sheet with each detail numbered. Note that the location and quantity of many of these details can be found in Figure 4.13. (Courtesy Jack Leaman, landscape architect, Ames, Iowa)

tural" members: each piece of wood is rated at a minimal characteristic strength in bending.

In addition to lumber grades, a few terms may be useful in the specification to clarify the exact quality or appearance of wood or the final product's appearance:

a. *Clear* indicates that knots are not acceptable in the finish work; it must be clear of all defects and blemishes.

b. *Kiln dry* indicates wood that is artificially reduced to a moisture content less than 12%.

c. *Seasoned* indicates wood that is naturally air dried to between 12% and 17% moisture.

d. *Sapwood* is wood that is usually lighter in color than heartwood of the same species, less resistant to decay, and cut from outside of the heartwood area.

e. *Heartwood* is wood cut from the center portion of the tree, which is usually darker in color than the same species of sapwood and often more resistant to decay than the sapwood.

f. *Vertical grain or edge grain* means that annual rings form an angle of 45 degrees or more with the face.

g. *Flat grain or slash grain* means that annual rings form an angle of less than 45 degrees with the face.

h. *Defect* is any irregularity found in or on a piece of wood that detracts from its appearance or lessens its strength.

i. *Blemish* are mars, scratches, gouges, mill marks, bark remnants, pith, discolorations, and gum pockets that degrade appearance and sometimes strength.

3. **Texture.** A designer must specify the nature of a wood member's milled finish. Generally this decision will involve textural qualities, avoidance of splinters, cost, and the product's final surface treatment.

a. *Dressed.* Lumber surface is smoothed by plane and milling.

b. *Sanding.* Lumber surface is smoothed by various grades of sandpapers to a variety of degrees of smoothness.

c. *Rough.* Lumber is not dressed, bears saw marks, is of greater dimension than dressed lumber, and is near nominal size.

d. *Resawn.* Lumber is dressed by resawing rough lumber, which usually imparts semirough texture.

e. *Textured.* Various special and specified treatments that impart a relief texture, include sand blasting, etching, and brushing.

f. *S4S.* All four surfaces are dressed or smoothed.

g. *S2S.* Both faces are dressed or smoothed.

h. *S1S.* One face is dressed or smoothed.

i. *Surfaced.* Both faces are dressed or smoothed.

j. *Jointed.* Both edges are dressed or smoothed.

k. *Miter.* These are 90-degree corners made with fitted 45-degree edges, or a geometric combination of similar proportions.

l. *Ease.* Edges are slightly rounded, about $\frac{1}{8}$ inch in radius.

m. *Chamfer or bevel.* Edges are cut at approximately a 45-degree angle to the edge and face.

4. **Sizing.** Although several schools of thought exist, it is generally best to size all wood members consistently on either enlarged plans, sections, and/or elevations rather than on details or the written specifications. Schedules may be used only if it is clear which member is of a particular size. Sizing on a plan, a schedule, and/or details is redundant and subject to inadvertent errors.

Nominal Dimension

Construction materials often vary in their dimensions to such a degree that a nominal—that is, an average or estimate—is commonly specified in place of actual sizes. A nominal size is a shorthand, convenient, and speedy means of expressing a dimension. Behind each nominal dimension is an implied understanding that only an estimate is being communicated.

Some materials, such as concrete block, brick, or pavers, are nominally expressed by three whole numbers in a specific sequence. For example, one concrete block is nominally an 8 × 8 × 16-inch size—that is, 8 inches high by 8 inches wide by 16 inches in length. The same block will be manufactured to actual dimentions of $7\frac{5}{8} \times 7\frac{5}{8} \times 15\frac{5}{8}$ inches.

Yard lumber is usually expressed in two dimensions, such as 2 × 4 or 2 × 6. Sequentially the wood is nominally 2 inches thick and 4 inches across the face width. The same piece is standardized by the National Grading Rules for Softwood Dimension Lumber, at a seasoned state, to average an actual size of $1\frac{1}{2} \times 3\frac{1}{2}$ inches.

Pipe and pipe fittings are also specified in terms of their approximate diameter rather than in their more complicated and precise actual inside and outside dimensions. The use of these dimensions is discussed in a following chapter.

Several basic rules should be followed when deciding how to coordinate material dimensions. Unless consistent decisions are made throughout both drawings and specifications, there can be communication of both redundant and conflicting information.

The following concepts will help to avoid confusion in documents.

1. Nominal dimensions are normally written on drawings and can be used in formal as well as casual conversation.

2. Nominal dimensions can commonly and implicitly function as a *complete dimensioning system*. Actual dimensions are defined in the technical specifications, that is, a nominal dimension is defined as to its actual dimension.

3. Care should be taken not to mix dimensional systems inadvertently. For instance, once a nominal dimension is noted on a drawing, there is no need also to delineate running dimensions describing the same object's actual or nominal size.

4. Once a manufactured product is specified, it will implicitly and automatically be dimensioned by the manufacturer's technical data. An attempt by the drafter independently to arrive at a nominal dimension can run counter to the nomenclature of both contractor and manufacturer. In general, the specification of a manufactured product carries, with the act, dimensional information by implied reference.

5. The overall site or detail dimensions should be double checked for conformance with the nominal or actual size of component construction materials—that is, do the small component units fit perfectly or imperfectly into an overall dimension?

6. Details should reflect a material's actual dimension, if the scale of the drawing is large enough to make such dimension relative; otherwise nominal dimensions can be used.

Dimensioning

The nominal dimensions of yard lumber produce some "rules" for dimensioning drawings and specifying quantities that must be recognized. For example:

1. Wherever possible, dimensions are run from center to center among several members. It is not possible mathematically to include either nominal or actual dimensions in a sum of wood dimensions because inaccuracies will result.

2. Dimensions may be run between two wood members, that is, face to face or outside to outside dimensions.

3. Unless a designer intends that precise cutting and dressing take place, only common sizes of yard lumber edge, face, and length should be specified. Ordinary usage calls for the sequence of edge and face to be written in nominal dimension and length as actual. For instance, $2 \times 4 \times 16$ indicates a wood member that is a nominal 2 inches on edge, and 4 inches in width, and is cut precisely 16 feet long.

4. Nominal wood members that are "ripped" (split) on site by a contractor will not meet standardized actual dimensions. For instance, a 2×6 should have an actual face dimension of $5\frac{1}{2}$ inches when dressed. When ripped into two members, they each will be approximately $2\frac{3}{4}$ inches wide, exceeding the "actual" dimension of a nominal 2×3 by $\frac{1}{4}$ inch. The specifier may wish to control on-site ripping of wood if variable dimensions may be a problem.

5. If any wood members vary in the shape from the yard lumber shape—that is, the common rectangular cross section—the specifications and drawings should recognize this. For example, each wood association and ordinary woodworking recognize various sectional shapes by common terminology and product literature. If the desired shapes are not ordinarily milled, it may be necessary to include large-scale drawings of the item with full dimensions of the shape. The specification writer should ensure that either the contractor is notified of the need for a shop drawing and/or a sample product is submitted to the designer for approval prior to actual installation. The drafter should carefully identify the quantity of special shapes on the various drawings.

6. The term *workmanship* is often used in specifications in an attempt to identify the quality of

product produced by a contractor. All too often, the term is supplemented and defined only in terms of the designer's approval. However, it is the specification writer's responsibility to seek an objective measurement of the contractor's efforts and skills as well as to reduce the need for subjective approval. Many regions have unions, cabinet makers, and woodworking institutes that may serve as references in judging rough and finish workmanship quality. The specifier can use terms that may be checked, as per Section 4.6.3, with definitions pertaining to the finish product. In addition, such terms as plumb, vertical, flush, fit, joint, trim, tight, loose, swing, rotate, level, square, true to line, visibility of imperfections, and joint fillers might be defined as to acceptable quality or acceptable function as required by the structure. At issue here is the basis upon which a contractor's work is to be accepted—what is required with respect to workmanship and what is unacceptable? Do the specifications reflect a true understanding of a designer's expectations in light of ordinary or exceptional workmanship?

7. The following items of connectors and hardware should be specified:

a. *Nails.* Size and space detail or nailing schedule using penny (d) symbol. Shank, head, point, and material, if they are not "common." Local or uniform codes may substitute by reference.

b. *Wood screws.* Size and space on detail or fastening schedule by diameter number and length. Type of head and whether slotted or Philips. Material with standard threads and need for pilot hole. Relate countersetting.

c. *Lag bolts* (lag screws). Size and space on detail or fastening schedule by diameter and length, with washers beneath head. Galvanized or ungalvanized finish with standard threads. Size of pilot hole and whether head is to be countersunk.

d. *Bolts.* Size and space on detail or bolt schedule by diameter and length, with or without washers beneath head and/or nut. Most light construction bolts will be standard steel rather than high strength. Galvanized or ungalvanized finish with standard threads and number of threads visible after tightening of nut. Whether head, nut, and washer are to receive any rustproofing with paint after installation. Pilot hole undersize and any countersinking.

e. *Anchors.* For fasteners installed into existing or proposed hard materials acceptable shield materials and installation practices (nails, machine bolts, and screws require different anchor or shield materials and design). Location, size, type, and pilot hole size, but not procedure (contractor should determine, for example, how to locate and when to install).

f. *Options.* Rather than specify the exact dimensions of wood fasteners, it is possible to write the specifications in performance terminology. Such specification language would contain minimum acceptable relationships among the forces to be received by a joint, minimum penetration of nails or screws, the maximum number of nails or screws per board size, maximum allowable withdrawal resistance per member, and similar information in a schedule. In effect, the contractor is left with the decisions regarding exactly how each connection is to be fastened. However, little control over a joint's appearance and a confused structural responsibility can result from such an approach.

Local codes often reflect the minimal performance of various fasteners under various stresses and uses, and should be used as related to permits and construction inspection by local authorities. Many of these codes, however, are rightfully concerned with structural integrity or ordinary frame construction and do not recognize nonresidential, esthetic, or unusual forms of privately constructed exterior structures.

8. If hardware is to be "off the shelf/standard," the terminology of drawing and specifications should be matched and manufacturer, address, catalog number, material, and common terminology of the piece specified. If hardware is to be fabricated, forged, or cast, the contractor will require large-scale detail drawings of the piece indicating the physical dimensions, material, thickness, finish, machining, function, and context of its installation. A designer may wish to include an "allowance" in the agreement and/or specify limited physical dimensions and function with precise shop drawings supplied by the contractor for designer approval.

It may be necessary to specify the finish condition and workmanship of the installation.

9. Two distinct methods of wood preservation exist and must be recognized in the specifications:

a. Painted or dipped application methods involve the definition of materials that may be used on site by a contractor and how the material is to be applied. In many instances, the timing, before or after construction, will be defined by the specifier. Generally there is no objective test for compliance, other than ensuring that specified or manufacturers' methods were followed by the contractor. Materials may be specified by reference to specific manufacturers or the common name of a material (creosote, preservative paints).

b. Pressure-treated wood may be specified as to the material to be used and the treatment process. For example, standards for the control of material quality and treatment procedures may be related to the American Wood Preservers Association (AWPA) or similar trade associations. In many instances, specifications of the various trade associations will include federal specifications, materials quality, methods of testing compliance, and the amount of preservatives to be held in the wood after treatment. In addition, the specifier should relate the treatment of the wood after construction—painting or dipping cut ends, damaged surfaces. The preservation specifications should be related to drawings that identify specific portions or members that are to be treated.

Specifications also should be related to the grade of wood. Is there potential conflict with decay-resistant heartwood species and a possibility of redundant preservation?

10. Specification of wood products differs appreciably from that of ordinary yard lumber. As a product, plywood, for example, possesses dimensions and dimensional stability more or less consistent with published data, that is, its "actual" dimensions are reasonably precise. Because the various products are manufactured, it is common practice to reference the material quality, performance characteristics, and, in many instances, installation, preservation, and construction techniques to manufacturer-prepared specifications and details. Materials will include plywood, laminated wood members, particle boards, and hardboards. In most instances, following the manufacturers' association specifications will assure both federal and regional code conformance.

Specifications should include a drawing's physical portrayal of the various patterns and arrangements of panel-like sheets, precise thickness and number of plies, the characteristics of the bonding agents or whether the material is exterior or interior grade, surface (face) appearance or texture, nailing or local code nailing schedule included or refer-

enced, edge treatments (plain, tongue and groove), and structural strength requirements.

Laminated wood members for light construction will usually involve appearance-graded members that are stock dimensions rather than special forms. Specification information will include the need for shop details for timber connectors; manufacture; standards of manufacture, such as the manufacturer's product association; allowable stress values within the member; conditions of use, either exterior or interior; appearance quality; finish coatings, manufacturer wrapping or other protection during shipping; guarantee or certification of manufactured quality; preservation treatment; provision for hardware attachment; member size in width, depth, length, and shape; and range of wood species used.

11. Specifications should control the quality and manufacturer of finish materials, such as paint or stain. Drawings locate finish type and color locations on each view or per a schedule. Testing may be limited to control of the number of coats applied or else a performance based upon applied mil thickness.

4.7 CONCRETE STRUCTURES

Concrete for site work is used mainly for paving, walls, retaining walls, and occasionally for lids over open boxes, and for various cast-in-place or precast drainage structures. Primary considerations in detailing and construction specifications involve whether or not the constructed item is to be poured in place or precast off site and installed on site.

4.7.1 Poured-in-Place Concrete

Pour-in-place (in situ) concrete necessitates complete detailing of a structure's physical dimensions and internal material characteristics. Because concrete is a plastic material, the designer is, in reality, dimensioning the forms in which the concrete will be placed. A designer may specify precisely the finish product's dimensions and trust to a contractor's ability to construct forms that will reasonably produce the desired dimension. It is a designer's responsibility to dimension the final concrete product, specify the quality of materials or necessary strength of the concrete, and describe the surface colors and texture desired.

Two methods are ordinarily used to control the quality of concrete within the language of technical specifications. They must be mutually exclusive if conflicts among guarantees are to be minimized. The first method can be referred to as *procedural* in nature. A designer controls the concrete's quality and strength by specifying the quality of the ingredients, their proportions, and procedures for placing and curing. By writing procedural specifications, a designer has accepted responsibility for the concrete's strength. A contractor accepts responsibility for carrying out those procedures and mixtures specified. Many small site work projects have used this method of specifying for years without problems. Even if specification errors should exist, both designers and contractors are aware of the approximate strength of concrete resulting from various proportions of materials. Errors can be found and corrected quite easily if everyone is accessible and cooperative.

A second method might be referred to as *performance* in nature. The technical specifications' language seeks objectively to control the end product of the contractor's work but does not specify the methods or procedures to achieve that product. For example, specification language will fix the compressive strength of cured concrete within prescribed testing procedures. It is assumed that a contractor will design the mixture qualities and placement to meet the specified tests. Performance-type specifications are not limited to large quantities of structural concrete but the expense of testing often precludes their use for small work.

Unfortunately many offices prepare technical specifications that are a mixture of procedural and performance techniques. A procedural specification may conflict with a performance language, making it difficult to fix responsibility. For example, if a concrete mixture should fail a performance test and yet was proportioned and placed according to specific procedures, who is responsible for the failure? In general, courts seem to side with the contractor who follows the procedures outlined in the technical specifications, assuming, of course, that the contractor did, in fact, follow the procedures prescribed. The final verdict will generally argue that a contractor is bound to follow the outlined procedures and, therefore, in a case such as this it would be impossible to attain the required performance quality. A contract that cannot be carried out becomes void.

4.7.2 Procedural Specifications

Procedural specifications will include the quality and quantities of the following items.

Cement

Cement should be specified by manufacturer, by type number (as normal portland cement type I), and/or by chemical ingredients; and also by reference to ASTM number (as ASTM C-150, type I) or other commercial specifications and various federal specifications.

Aggregates

These should be specified by size as percent passing various screens or acceptable graded range in millimeters of diameter, by soundness as a measure of density and weight, by reference to an ASTM number (as ASTM C-33 for concrete aggregates), or by local graded sources that may be commonly used. Local highway departments are good sources of aggregate specifications. Acceptability of crushed or gravel materials should be specified, as well as the maximum size of aggregate relative to the size of reinforcing steel and distance between forms. Designers must recognize that local availability is necessary.

Water

Although specifications may list a variety of foreign matter not acceptable for mixing with cement and aggregates, the single word *potable* can usually serve as a specification—for example, "Water is to be of potable quality." The statement implies that the water must be of drinkable quality. However, if a mixing site must use locally available nonpotable waters containing foreign matter or that produce chemical reactions, it may be necessary to specify the amount of oils, organic matter, silt, acids, alkali, and other deleterious matter that is acceptable. Specifications should clarify responsibilities if the water is owner supplied.

Admixtures

The manufacturer, material, and proportions of any admixtures, such as lime or air entrainments, must be specified. Generally the contractor is cautioned to follow the manufacturer's directions as to mixing and placement.

Proportions

The very nature of procedural specifications implies designer control of the proportions of ingredients that are combined and the methods of mixing and placing the material. Proportions are generally indicated by a three- or four-numeral sequence. For example, a proportion of $1:2:3\frac{1}{2}$ commonly indicates that one part cement, two parts sand, and $3\frac{1}{2}$ parts gravel are to be combined and mixed. The numbers are always in that sequence. In a mix specified as $1:2:3\frac{1}{2}:\frac{1}{4}$, one-fourth part is an admixture also to be combined and mixed. The contractor then looks for a specification of the admixture material. The water portion of a mixture is ordinarily specified in terms of gallons per sack of cement and as including free water existing on the surface of aggregates.

Specifying a mixture combination as proportional implies on-site volume portions, that is, one shovel of cement, two shovels of sand, and so on. Obviously there can be variance in measurements of the mix. If a designer wishes more precise control over measurement, this should be specified. Proportional mixtures are understood to be less dependent upon a contractor's measurements and may result in a range of compressive strengths.

Tests

Proportional mixes are difficult to test for anything except a contractor's adherence to prescribed procedures, visual analysis of what constitutes ordinary quality of workmanship, slump tests, and a reasonable flexibility to make minor changes in the mix as situations develop. A simple slump test can be specified to verify the amount of water in the mixture. For example, specifying a maximum slump of 3 inches and a minimum slump of 2 inches will give some control over the concrete's consistency during placement and a degree of objective measurement of the mixture's water content and probable strength. A contractor may then make minor adjustments to a mixture's constituents to equal the slump tests's permissible range.

Mixing

Both procedure and equipment should be stipulated for the mixing of proportionally specified concrete. Generally the time of mixing, the type of equipment to be used, and the rotational speed of the mixing drum, if one is used, are specified. It is common practice to allow a contractor control over the mixing place—on or off site. However, if mixing is done off site, the specifications must control the maximum permissible time between the addition of water and the concrete's on-site placement as well as the use of tempering water once a concrete arrives on site. If mobile mixing trucks are used by the contractor, a weight and proportional certification slip should accompany each delivery.

Cleanup

The chemical and physical nature of concrete makes its on-site deposit during cleaning of transit mixing truck drums and on-site mixers critical to plant material life as well as the work of other contractors. A contractor should be directed where to deposit excess uncured concrete or wash water during construction. Excessive deposits of concrete, plaster, and their wash waters will prevent some plant growth because of the presence of soluble salts in the form of sulfates and carbonates of sodium and potassium. Soil may be made severely limiting to plant growth by the casting of lime and cement from broken sacks or containers while stored on site. Many admixtures contain sodium, calcium, oil, plastics, and other chemical bases that may damage or sterilize soils during construction operations. If such problems are considered critical, the specifications should direct storage, cleanup, and removal of contaminated soils prior to planting operations.

Curing of Concrete in Place

Although the proportional method of specifying concrete does not always lend itself to precise control of a mixture, it is generally recognized that specifications will control the curing process. For instance, technical specifications can direct certain operations and alternative methods to protect concrete during hot and cold weather. Such specifications will classify the protective measures in terms of temperatures. Moisture control is generally the prime reason for directing the curing process. If chemical compounds are to be used, they should be specified and their use can be directed to manufacturers' specifications by reference.

Placement

It is possible but cumbersome to include the do's and don'ts of placing concrete in technical specifications. Referencing to placement directives for ordinary concrete work published by the Portland Cement Association or the American Concrete Institute will reduce the complexity of technical specifications and precisely control the placement procedures. However, it is not out of the ordinary to control local workmanship in the placement of concrete by identifying key procedures in the technical specifications.

Mix Design

For work that requires precise control of the proportions and strength of a concrete mix, the owner may wish to employ a mix design and testing system. The specifications will usually contain an outline of the material quality and size limitations. By using the specified materials, an engineer or laboratory develops precise combinations of materials to produce a cured concrete of specified strength. The contractor is responsible for the correct proportioning of the materials as directed by test reports. Assuming that a contractor's efforts in proportioning and placement are correct, an owner and designer are assured of a cured concrete that will be within 15% of the specified design strength. A contractor agrees to carry out the procedures and proportions as indicated by laboratory tests. Samples of the mixture are tested during specified periods and in a specified ratio to the volume mixed. If the laboratory finds that a mix is not achieving a desired strength, the contractor will be directed to modify either proportions or procedures. Specifications must clarify what will happen if the specified materials, procedures, and designed mixture do not meet the specified strength.

4.7.3   Performance Specifications

If performance specifications are carried to their logical end, a designer might simply delineate the scope of a finish product, characterize the esthetic and structural quality of the finish product, and allow a contractor to assume responsibility for achieving that product. Theoretically that is how a performance specification functions, and it often does work that way on small projects. For example, many small portions of site work are executed each day by owners and designers who explain verbally what they want and a contractor carries out the job.

Unfortunately the process is not always successful when the parties fail to explain exactly what is desired or acceptable.

Most major projects depend upon a combination of procedural and performance specifications, even though the responsibilities may be somewhat clouded in the process. Designers are probably not willing completely to relinquish control by writing performance specifications in all cases.

For concrete work, the performance specifications may be written so as to place total control in the hands of a contractor. In essence, a designer is allowing a contractor to make all decisions regarding concrete proportioning and placement. For instance, swimming pools, poured concrete retaining walls, and off-site precast structural members or elements are often specified under performance concepts.

Concrete lends itself to performance concepts because the finished product can be judged objectively, that is, the work can be judged against a set of criteria and accepted on the basis of the work's ability to meet those criteria.

Such criteria include *compressive strength*. Concrete is tested to meet or exceed a certain pressure per unit of concrete within a prescribed number of days and within prescribed testing procedures. For example, typical language might be: "Minimum allowable compressive strength shall be 4000 psi at 28 days for typical specimens for each 200 yards of placed material taken in accordance with AASHTO T-23 and tested in accordance with AASHTO T-97." Additionally, the payment for testing and for failure of tests is specified.

Taken alone, testing for structural strength would not constitute a complete specification—but it would adequately control and guarantee the structural soundness and quality of the finish product. Going even farther, core samples can be taken from the cured work and measured and tested for in situ compliance with criteria. The methods, materials, placement, and procedures necessary to meet the criteria are at the contractor's discretion. Very often this sort of procedure is used by design/build or turnkey firms where an owner can receive protection by hiring an independent laboratory to test a contracting firm's compliance with quality guarantees.

In addition to performance characteristics relative to strength, a designer should delineate and specify the pattern of joint placement, texture, color, patching and repair, and subsurface compaction.

4.7.4   Special Needs

Joints

Concrete specifications define the various joints and drawings detail their dimensions and character. Expansion joints should provide for filler material to seal water from the subbase, using the manufacturer's data or ASTM references. Construction joints either may be formed or cut as directed. The construction site plan should be coordinated with various joints in slabs on grade, and in curbs, walls, and detail sections. Particular note must be made when joints in slabs on grade must coordinate with a module, exterior furnishings, or architectural patterns. If necessary, an enlarged plan view of the slab on grade should detail precise locations of all joints, including cold joints.

Texture

Vertical concrete surfaces are usually textured by specifying the characteristics of containment forms. For example, planks and plywood will be reflected in the cured concrete's texture when used as form materials. Semicured vertical surfaces can also be textured by the choice of finishing tools, application of matter during the curing process, exposure of aggregate during the curing process, or mechanically imprinting patterns. The list of techniques is endless and cannot be amplified here.

Vertical surfaces require specification of form material as it affects the surface texture. The choice of form structural design is ordinarily left to the contractor. If necessary, a detail drawing should be prepared that graphically portrays the texture sought, any dimensions to a relief or pattern, and suggested means of accomplishing the texture. However, it is common practice to allow a contractor considerable latitude in developing procedures necessary to accomplish a desired texture. Within this context, specification language may allow latitude in the use of concrete retarders, form coatings, accelerators, or other means of manipulating the timing of a concrete's curing process while it is in the forms. It is always best for specifications to direct a contractor in preparing and submitting physical samples of a texture for designer approval prior to full-scale construction of forms.

Horizontal surfaces require that textures be described, and sometimes procedures outlined, in specifications. If a particular texture is common locally, it may be possible to trust contractor experience. However, the safest thing to do is to describe it, outline procedures, and require the contractor to submit a sample. Once approved, a sample is stored for later use as a comparison and means of evaluating the contractor's workmanship.

Color

Concrete to be colored is very similar to texturing inasmuch as a sample may be the best solution. A designer-approved sample allows a certain degree of latitude in the choice of coloring materials and methods of construction. However, an experienced designer will usually specify the manufacturer of the coloring agent, the color number, the type of cement necessary to achieve the color, necessary proportions of the coloring agent in the concrete mixture or applied to the concrete's surface, uniformity of the soil subbase and aggregates, use of "topping" mixes containing coloring agents, curing of the concrete, and a sample of the contractor's work and color achieved. In some instances, the cured concrete may require the application of protective hardeners or other maintenance. The designer should coordinate the colored concrete with any texturing and quality of form work for vertical surfaces, and the gradient of slabs on grade, if retention of drainage water is necessary for proper curing of the surface.

Compaction

Coordination of utility or other excavations beneath slabs on grade is necessary to prevent subsidence of soil and slabs. Specifications for backfills must direct mechanical compaction to that density required of all structural fill areas.

Patching and Repair

Most concrete work will require some degree of correction of imperfections in surfaces. Some textures may require little patching whereas very smooth and visible surfaces will require careful control of the repair material's color and permanency. Colored or smooth concretes with imperfections may require cutting and replacement of portions. In any event, the specification writer will control this aspect of the project.

Defects in the concrete work may be structural problems as well as appearance conditions. A contractor should be directed as to corrective measures and the designer's and contractor's responsibilities with regard to structural defects.

4.8 MODULAR MATERIALS

Brick, concrete block, wood, adobe blocks, and tile are common elements of site work. Modular units are distinctive in the sense that drawings and specifications must be coordinated in a specific unit size. For example, a dimensioning plan should locate objects in the modular unit size of a respective material, and the specifications must key to the quality, size, and installation work.

Modular units are much like wood with regard to the use of nominal and actual dimensions. Each unit is manufactured and, except for adobe, has a fairly uniform and accurate physical dimension. However, the actual dimension is smaller than the nominal dimension. It is assumed that each module will be installed with a joint or space between each unit. Dimensioning techniques must recognize the existence of such a space by dimensioning to the center of that space.

A designer must decide, delineate, and specify answers to each of the options noted in Figure 4.21 when preparing contract documents: (1) grade of brick or block, sizes in nominal dimension, and pattern of each units arrangement; (2) characteristics of mortar, striking of joints, joint thickness, metal ties to any facia units, and necessary bolt setting; (3) size, spacing, and characteristics of

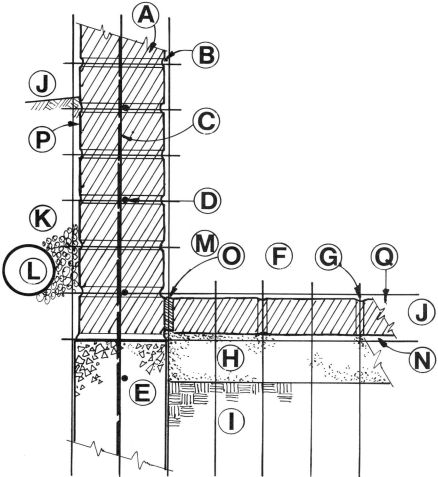

Figure 4.21 Typical sectional detail of modular unit construction. Note that the modular units are drawn within a 4 × 4-inch module.

vertical reinforcement, grouting, setting of bolts, location of any pilasters; (4) size, spacing, and characteristics of horizontal reinforcement or cross-tieing; (5) size, material, reinforcement, and depth of footing below grade; (6) grade of brick, block or other paving material in nominal sizes and patterns; (7) mortar or butt joint in paving and form of striking; (8) aggregate or structural subbase material, depth, and characteristics; (9) compaction of subsoil, responsibility for grading; (10) relation of modular size to height and length of a wall or paving pattern, spot elevations, detail dimensions, and finish elevation of surrounding soil; (11) character, type, size, and quantity of silt filter around pipe or weep hole; (12) type and size of any subdrain pipe, location of invert elevations and discharge points on storm drain plan; (13) character, size, and spacing of any weep holes; (14) weed control membrane or treatment specified and located; (15) expansion joints as required; (16) waterproofing as necessary; (17) detail edge finish or containment of modular edges.

4.8.1 Brick

Brick is a material with variable precision as to its actual dimensions, composition, and quality. Drawings should graphically define the nominal size, space between units, pattern or placement, location of mortar, and configuration of the mortar joint (striking off). In addition, walls and footings are usually given an elevation on the grading plan and located on the horizontal control or enlarged detail plan; cross sections and elevations indicate footings, footing steps, reinforcement requirements, and cap design. Spot elevations should be coordinated with vertical dimensions on details (a common mistake is that spot elevations do not agree with vertical detail dimensions or with the nominal dimension of the brick and joint). When used as paving, the subbase and brick should be detailed in cross section with edge treatment shown. The grading plan will indicate spot elevations necessary for storm water drainage. When brick is used as a facia, the brick wall ties, type, and spacing should be delineated.

Specifications will include the type of brick to be used, its nominal or actual size requirements, mortar type and characteristics, workmanship quality, grout quality, and reinforcement within the mortar course and vertically.

Brick is referenced, like wood, in its nominal dimension. For example, a "common" brick might

be noted as being 4 × 3 × 8 inches nominally, but can be about $3\frac{3}{4} \times 2\frac{1}{2} \times 7\frac{1}{2}$ inches in average dimension. With a $\frac{1}{2}$-inch mortar joint, the finished dimension will be 4 × 3 × 8 inches. The finish dimension then depends upon the locally available actual dimension and the dimension of the joint.

Both design and quality of materials must recognize that brick for exterior use will be exposed to greater degrees of weathering and climate than is commonly true of architectural walls or interior wall faces. Bricks must be of highest quality, mortar must be medium strength for flexibility during expansion and contraction periods, vertical expansion and contraction joints must be detailed, brick must be protected from continuous absorption of water, and freezing and thaw cycles must influence the manner of joint striking. Both brick and mortars should be free of calcium sulph-aluminate, iron particles or sulfates, and other chemicals that may produce efflorescence or deterioration.

4.8.2 Concrete Block Units

Concrete blocks are units precast on the basis of nominal size dimensions and various open cell patterns, face textures, colors, weights, and finishes. Because of transportation costs, a designer is usually limited to block units that are manufactured locally.

Four basic data are necessary to the specifications. First is the block's *appearance*, relative to the manufacturer and available patterns, face textures, colors, and finishes. Usually a manufacturer is named as supplier and alternate manufacturers of a similar product are also mentioned. The *size* of each block is specified in nominal terms with the usual designation sequence as, for example, 8 × 4 × 16 (width × height × length). As with brick, the ultimate height of walls or the width of paving will depend upon the width of a mortar joint. *Structural characteristics* of concrete block generally vary in the same manner as normal concrete varies with the water : cement ratio, and aggregate characteristics. Hollow units must be designated as to their load bearing characteristics by grade and ASTM designation.

Mortar can be specified by proportion or ASTM designation, along with whatever plasticizers or workability admixtures are required. Grouting is usually required by the presence of reinforcing steel or to fill otherwise hollow cores for strength. The number of cores to be filled and the physical

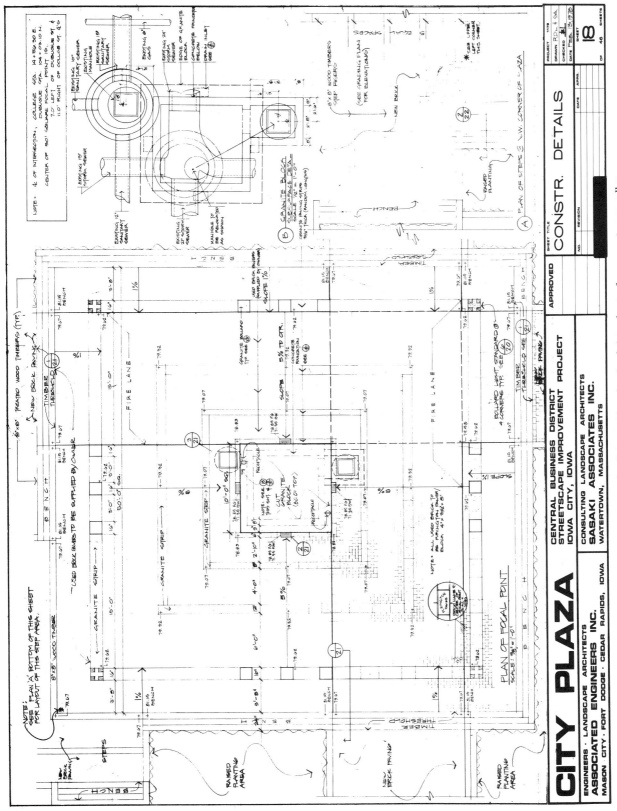

Figure 4.22 Typical construction detail sheet with each a simple intrasheet reference system as well as reference symbols to other sheets in the set. Note that this "plan of focal point" is centrally located on Figure 4.13 and is an enlarged detail plan necessary for the precise identification of special construction details. (Courtesy Jack Leaman, landscape architect, Ames, Iowa)

characteristics of the grouting should be specified.

Concrete block may be used with facia brick, plaster, and the like, and may require specification of facing ties, horizontal tie material within each mortar joint, or stucco wire mesh applied vertically to the block's face. Each of these operations will require both drawings and specifications.

Walls of concrete block may require specification of waterproof membranes, paints, emulsions, or liners as necessary to the design.

Visible finish surfaces will require attention to the finish of joints, that is, tooling and the form of the joint. Cleaning and painting may be necessary.

Concrete block details and spot elevations require coordination in that either spot elevations or details should control the vertical dimensions and relate the nominal and actual dimensioning to the joint width. Horizontal controls are usually applied to the face of concrete block. Checks should be made for redundancy and errors among detail dimensions, nominal designations of block size, written specifications of materials, and site construction plan dimensions. Width and depth of footings are to be coordinated with plantings, excavations, and utilities, as well as with existing conditions.

4.8.3   Tile

For tile, material and manufacturer should be specified with regard to specific nominal size, color, and finish. Drawings should indicate cross-sectional detail, arrangement, joints, spot elevations, overall dimensions, setting bed thickness, reinforcement, and adjacent materials. In some instances, the joints may require sealants or special adhesive setting materials to resist water penetration.

Mortars for exterior work must be strong, yet flexible enough to take the extensive expansion and contraction or heaving that often affects walls and pavings. Guide specifications are available from the Brick Institute of America, the Portland Cement Institute, and the Structural Clay Products Institute. Mortars are generally specified as to their type—specific proportions of mix constituents, water content, admixtures, and preparations. Each type is identified in, for example, ASTM designations for quality of materials and testing procedures. Each type of mortar must be selected for strength, resistance to weathering and chemical conditions, workability, and suitability for the design.

4.9   MANUFACTURED PRODUCTS

The context and understanding of responsibilities regarding the "or equal" clause, "open" and "closed" specification language, and "as approved" procedures often afford contracting parties some degree of confusion. In brief, an "or equal" clause allows a contractor to suggest products or materials as substitutions for that which is specified and to be approved by a designer. The term *equal* must be defined in terms of equal function, quality, quantity, cost, longevity, performance, and so on. The term *as approved* requires definition as to who approves, procedures for approval or disapproval, and potential appeals upon decisions. *Open* language is, by its nature, a performance orientation that allows a contractor total, or almost total, selection of products and materials. *Closed* language is oriented to a single or very limited manufactured product or material. *Open* language allows many contractor choices within the bounds of the specifications whereas closed language limits choices to those explicitly mentioned in the specifications. An "or equal" clause may wedge open closed language if approved by a designer.

The interrelationships among various clauses and meaning of language affect the manner in which drawings and construction bids are prepared, and the clarity of contract administration during the progress of site work. The pros and cons relative to this subject are as varied and complex as are the designers, owners, contractors, manufacturers, guarantees, and construction and purchasing law. The following notes examine important concepts and interrelationships but by no means fully explore this complex subject.

1.   In essence, a closed specification implies that, in the designer's opinion, only those products specified are suitable for use in the contract work. Any substitution by the owner or contractor must be approved by the designer.

2.   Many public works projects do not ordinarily allow closed specifications. However, a designer may, with valid reason, write a closed specification for a specific product or material with the understanding that both contractors and manufacturers may wish to question the validity of reasoning.

3.   Any closed type of product specification must be coordinated with any type of performance or product guarantee.

4.   Drawings and details should not identify a

product by firm or product name, but should reference by generic or construction terminology. For example, if a drinking fountain is shown, it should be identified as a drinking fountain or as drinking fountain A, B, or C. The manufacture of the drinking fountain(s) is identified in the technical specifications. In this way, should substitutions occur during construction, a record of manufacturer(s) will be maintained in the change orders and specifications. If a change order is written, drawings do not have to be revised as the change order records the revisions and the drawings stand unchanged.

5. If a designer knows of several manufacturers who produce equivalent products, these should be written into the specifications. Such a habit allows a contractor greater flexibility as to sources and saves telephone calls and paperwork.

6. Written paragraphs concerning guarantees and operational information are to be correlated with those products specified. For example, a control valve for irrigation may carry a written guarantee that differs from the length of time for which the entire system is guaranteed by a contractor. If the valve is guaranteed for two years and a contractor's guarantee expires after one year, the owner should have such knowledge for maintenance and replacement purposes. For many manufactured items, maintenance, operational, and accessory information is available from the manufacturer. Such information should be assembled by the contractor and delivered to the owner or designer prior to final payment for the work.

In this same area, a designer might be advised to direct an owner to purchase certain publications by manufacturers that will aid in the maintenance and operation of various products.

7. Very often, a manufacturer will produce many specifications for a variety of products and it may not be possible to pin down exactly which specification is controlling the situation. On the other hand, a manufacturer may not have published any specifications or may revise specifications during the time of design, bidding, and construction. The question then becomes: Which specification controls? Many manufacturer guide specifications are intended for reproduction or reference in a designer's specifications. They may be of the "fill in the blank" type, with certain decisions necessarily left to the designer. Unless the designer fills in those blanks, the information may be quite useless as a technical specification.

8. Some manufacturers guarantee their product or materials only if installation procedures follow their specifications or personal directions. Failure of a designer to recognize this may nullify the guarantee. A few product guarantees are based entirely upon its installation by the manufacturer's own personnel. When this situation occurs, the prime contractor must be made aware of it and the work coordinated in the special conditions, bid form, and agreement for construction.

9. If a designer modifies or changes a product in some fashion, the manufacturer's guarantee and responsibility may be clouded. The manufacturer must be consulted in all matters pertaining to a product's use and modifications of abnormal character. Particular concern must be given to personal liability concerning a product's use.

4.10 ASPHALTIC CONCRETE

Many regions of the United States must depend upon local batch plants for availability of asphaltic concrete or similar products. Such material is not ordinarily controlled by the contractor, that is, it cannot be mixed on site and so depends on off-site production. Technical specifications cannot change the fact that only certain materials are available.

Local highway department specifications can be consulted to determine local specifications that are generally followed by the local batch plant. Use of highway specifications will ensure local availability. Special asphalt construction, such as playgrounds, pool linings, and walkways, should be discussed with batch plant personnel to determine the quantity of materials required for special batching and the mix proportions necessary.

Generally asphaltic concrete will require a sectional detail of various layers with their respective generic terminology, subbase compaction, gravel subbase materials, surface sealers, spot elevations of gradients, edge confinement detail, asphaltic compaction needs, material qualities and proportions, and subsurface drainage needs. In some regions, it may be possible to use porous asphaltic concrete with special attention to the size and gradation of aggregates, surface gradient, and subsurface drainage.

Asphaltic concrete specifications follow guidelines prepared by the Asphalt Institute and ASTM designations.

4.11   SYNTHETIC SURFACES

In general, materials for synthetic surfaces will consist of relatively new plastic and asphaltic mats or emulsions suitable for the paving of exterior recreation or leisure use areas. Drawings will comprise the construction plan locations and detail cross sections. Specifications usually refer to one specific supplier and manufacturer of the products. All but small-scale residential uses will probably involve direct communication with the manufacturer and subsidiary companies for the material and installation. In many instances, the installation must follow the manufacturer's directions and be installed by a subsidiary company of the manufacturer. Very often, curing periods, temperatures, and sequential layers of materials will add extraordinary time to a construction contract. Before any agreements as to such considerations as contract time and liquidated damages are reached, the total time necessary for material curing and preparation should be checked.

4.11.1   Mat Materials

Specifications for mat materials will include complete chemical and physical control of the product and testing for manufactured compliance, installation of asphaltic concrete subbase, specification of a cushion beneath the surface mat, edge details with products often requiring a bond beam edge and securing of the mat to the beam, sewing or adhesion of the mat's edges to each other, painting and marking of the finish surface, and peripheral maintenance facilities such as water and electrical outlets.

4.11.2   Emulsion Materials

A family of synthetic materials is available for the sealing of asphalt and the development of recreational surfaces. Their use includes application as a liquid over old or new concrete or asphaltic concrete surfaces. Although the number of such products on the market makes a complete discussion impossible here, there are certain chemical and physical conditions that should be checked for each product.

1.   Many of these products require separate and distinct materials to be applied in various sequences and over various structural surfaces. Some materials are bonding agents only, and are necessary to prepare structural surfaces for the finish surface coating. Each type of bonding agent is matched to a specific type of paving and requires a specific number of days to cure. For example, it is suggested by manufacturers that new asphalt cure for at least two weeks and that new concrete age for 35 to 50 days prior to the application of a specific bonding agent. Old concrete surfaces may have to be sand blasted or patched and aged a similar length of time. The type of bonding agent required must be matched to the physical and chemical condition of the paving over which it is to be applied. The materials cannot be applied when the temperature is below 50°F or it is raining.

2.   Because of the foregoing conditions, the specified project time, and thus liquidated damages, must be fully examined before the construction agreement is signed. Construction time will not be normal but will run considerably longer as compared with normal concrete or asphalt paving construction.

3.   A cushion coat of material can be laid over the bonding surface and may consist of 4 to 10 coatings. Each of these must cure before the next is applied. The specifications must carefully define the number of coats and the constituents of the mixture.

4.   A finish or weathering surface will usually be applied over a cushion layer. The finish must be specified as to its asphaltic or acrylic emulsion characteristics, color, texture, and stripping materials.

5.   The product's guarantee must be matched to the technical specifications for installation procedures and curing times. A check should be made as to whether the specifications will meet or nullify the manufacturer's guarantee of materials. Does the guarantee in the construction agreement cover both materials and the finish surface? Does the manufacturer require any sort of on-site supervision or prior approvals of construction methods as a part of the guarantee?

SELECTED READINGS AND REFERENCES

Alpern, Andrew, Editor *Handbook of Speciality Elements in Architecture.* New York: McGraw-Hill, 1982.

Carpenter, Jot D., Editor *Handbook of Landscape Architectural Construction.* Washington, D.C.: American Society of Landscape Architects, 1976.

Ching, Francis D.K. *Building Construction Illustrated.* New York: Van Nostrand Reinhold, 1975.

Landphair, Harold C., and Fred Klatt, Jr. *Landscape Architecture Construction.* Amsterdam: Elsevier-North Holland, 1979.

Liebing, Ralph W., and Mimi Ford Paul *Architectural Working Drawings.* New York: Wiley, 1977.

Muller, Edward J. *Architectural Drawings and Light Construction,* 2nd ed. Englewood Cliffs, New Jersey: Prentice-Hall, 1976.

Stein, J. Stewart *Construction Glossary.* New York: Wiley, 1980.

Walker, Theodore D. *Site Design and Construction Detailing.* Washington, D.C.: American Society of Landscape Architects, 1976.

Chapter Five
Horizontal Control and Dimensions

In spite of great strides in perfecting instruments, 99 percent of what we design and build is accomplished with a rusty tape measure.

John M. Roberts

5.1 THE DRAWING AS INFORMATION

A contract document drawing transmits specific information to a contractor regarding the location and dimensions of all elements to be installed or constructed. Information is conveyed by a scaled drawing as well as by a variety of graphic dimensional systems and symbols that are unique to site work.

Site plan horizontal control is described by various sheet titles and terms relative to regional usage. For example, the plan may be referred to as a "staking plan," "layout plan," "dimensioning plan," or "plotting plan." Each of these terms describes a surveyor's or contractor's efforts in laying out, plotting, staking, and otherwise locating site work elements prior to construction.

From a designer's standpoint, horizontal control will include coordination among the various site plans, details, materials, sections, elevations, and technical specifications. As each of these documents develops, responsibilities will be given to each of the contracting parties through symbolism, implication, convention, and direct statements.

5.2 ACCURACY

Each site planning project may require several degrees of accuracy during construction and instal-

lation. One portion of a designer's responsibility is to identify just how accurate the work must be. Unfortunately many projects function with a designer making such decisions on site and without sufficient documentation of requirements to avoid opinionated conflict. Site work documents will vary from that accuracy necessary to fit small parts together to personal agreements and opinions relating to the location of an object in space. Some dimensions will require close survey work by registered surveyors whereas other work may be approximately located by a contractor using a rusting tape measure.

5.2.1 Opportunity and Need for Accuracy

Site work elements may vary considerably in their potential to be measured as well as to a designer's need for accuracy. Potential for measurement varies between naturally occurring materials and manufactured materials. For example, natural stone may vary so much that only an average range of sizes can be specified. A wall constructed of such stone may not be precisely located. Both the designer and contractor are saved frustration if the drawings clearly indicate and allow for any variation inherent in the materials used. On the other hand, manufactured materials are often assembled and located with only slight limits to accuracy.

5.2.2　Implied Accuracy

Accuracy may be implied symbolically and mathematically by the graphic and numerical presentation of data. Numerals fix a dimension as fact, yet the graphic manner of presentation implies the degree of accuracy acceptable to a designer. One example of this characteristic is a drafter's use of significant numbers. Significance is implied by the number of numerals written to the right of a decimal point or the fractionalization of a dimension. The designer essentially is saying, "I intend the contractor to be this accurate with measurements; as long as the measurements lie with the implied range, I will accept such work." (See Table 1.)

5.2.3　Fixed and Adjustable Dimensions

Site planning will involve a hierarchy of about four less-than-distinct areas of dimensioning concern. Each type of dimension will require attention to an acceptable degree of accuracy during document preparation and on-site layout. In general, each type is defined in terms of legal necessity or esthetic concern.

1. *Fixed construction* or lines must be computed and located to the highest order of surveying accuracy. For instance, property lines, baselines, rights-of-way, structures located on or about property lines, and conditions from which other

TABLE 1

Graphic Form as Given on Contract Drawings			Implied Average Range of Accuracy Acceptable for Ordinary Work		
Feet/Inches					
$58'$			$\pm\ 6''$		
$58'\text{-}1''$	or	$58'\text{-}0'$	$\pm\ \frac{1}{2}''$		
$58'\text{-}1'' \pm \frac{1}{8}''$		$58'\text{-}0' \pm \frac{1}{8}''$	$\pm\ \frac{1}{8}''$		
$58'\text{-}1\frac{1}{2}''$		$58'\text{-}0\frac{1}{2}''$	$\pm\ \frac{1}{4}''$		
$58'\text{-}1\frac{1}{4}''$		$58'\text{-}0\frac{1}{4}''$	$\pm\ \frac{1}{8}''$		
$58'\text{-}1\frac{1}{8}''$		$58'\text{-}0\frac{1}{8}''$	$\pm\ \frac{1}{16}''$		
$58'\text{-}1\frac{1}{16}''$		$58'\text{-}0\frac{1}{16}''$	$\pm\ \frac{1}{32}''$		
Feet/Decimal					
$58'$			$\pm\ 0.5'$	or	$\pm\ 6''$
$58.1'$	or	$58.0'$	$\pm\ 0.05'$		$\pm\ \frac{5}{8}''$
$58.11'$		$58.00'$	$\pm\ 0.01$		$\pm\ \frac{1}{8}''$
$58.111'$		$58.000'$	$\pm\ 0.005$		$\pm\ \frac{1}{16}''$
Angles					
$58°$			$\pm\ \frac{1}{2}°$	or	$\pm\ 30$ minutes
$58°\ 00'$			$\pm$ one minute		
$58°\ 00'\ 00''$			$\pm\ 30$ seconds		
Meters[a]					
58 m			$\pm\ .5$ m	$=$	$19.69''$
58.1 m	or	58.0 m	$\pm\ .05$ m	$=$	$1.97''$
58.11 m		58.00 m	$\pm\ .01$ m	$=$	$0.39''$
58.111 m		58.000 m	$\pm\ .005$ m	$=$	$0.197''$

[a]Dimensions in meters (m) are commonly given to at least three decimal places. Note that three decimal places produce a $\pm$ range of about $\frac{3}{8}$ inches and do not compare in accuracy to three decimal places in feet. Use millimeters (mm) for precise and detail dimensions.

dimensions may be measured must be fixed. Fixed elements either constitute a legal property right or serve as a basis upon which other dimensions will depend.

Mathematical calculations may be performed to derive a fixed dimension. Computed dimensional data may not be rounded off to less than three decimal places, and often extend to five places for precise legal evidence. Precautions must be taken during computation that precision is not sacrificed through an accumulative error by premature rounding of products.

2. *Semifixed dimensions* often locate an element of construction from a fixed construction point or line. Once a semifixed element is located, it may be used to locate other semifixed or floating elements. Very often, a semifixed dimension may be sufficiently accurate to avoid encroachment into setback zones, adjacent property, precise construction, underground obstructions, and the like, but not accurate enough for the legal recording of property. For example, an olympic-type swimming pool may be used for competition, but if it is constructed $\frac{1}{4}$ inch too short, it is unacceptable for competitive activities and is reduced to recreational status.

Semifixed dimensions constitute prime data for most site work. Accuracy is often desirable to two decimal places, or about $\frac{1}{8}$ inch, for most work. Elements so dimensioned are usually the first objects to be located on a site and form the framework upon which floating elements are located. A registered survey crew should be employed to achieve such accuracy. A designer must be aware of the extra cost in time and effort involved before specifying or implying such a degree of accuracy.

Semifixed dimensions often are a designer's decision rather than a legal necessity. Such dimensions thus exist somewhat arbitrarily within an esthetic design decision and may not have any basis in legal need. When semifixed dimensions are arbitrary, their computed products may be rounded to acceptable accuracy. Modern calculators may lead to mathematical products that appear to be correct to seven or nine decimal places. However, when the numbers entered are arbitrarily derived, then their computed products are not necessarily what they might appear to be. Generally the multiplication or division of data will produce garbage answers. For example, an esthetic design decision to construct a curve with an arbitrary 120 feet of radius and an arbitrary 60.2-degree included angle electronically

computes a curve length of 126.0825852 feet. The product should be rounded to 126.08, or even 126, feet to match the accuracy obtainable from the original arbitrary data. Of course, this assumes the curve to be arbitrary and not a legal property or right-of-way description. It is a disservice to everyone concerned if the drawings and contract require accuracy when the need for such accuracy is arbitrary.

3. *Adjustable or floating elements* often occur in site work whenever scaled dimensions or a great range of accuracy is otherwise specified. Only approximate locations are necessary or desired by both the designer and contractor. However, because many such dimensions are scaled, the drawing site plan becomes a critical element in developing whatever accuracy may be required by the designer. Obviously a drafter's inaccuracy, the weathering and changing dimension of print paper, and the difficulty of in-field scaling can lead to inaccuracy greater than intended.

Adjustable elements and dimensions are those that are located only after sufficient fixed and semifixed elements exist to provide a framework of positional reference. On occasion, adjustable elements may not be within a reasonable distance for taping or optical range-finding equipment measurement. Under such circumstances, topographic forms or other site features may be necessary as locational reference points. Examples of adjustable elements include plant materials, irrigation system components, lawn perimeters, loose paving areas, trails, and pathways through open or forested areas.

Adjustable elements require a mutual agreement among the owner, designer, and contractor. Specifications or the general conditions pertaining to the designer's role may establish the type of decisions to be made on site. Each technical specification should describe procedures, set responsibility for tentative locations, call for designer approvals as to locations, and explain the flexibility given to a designer to make minor adjustments to the contractor's tentative locations.

5.2.4 What's Fair?

Personal expectations will become involved in assigning fairness to horizontal control. A contractor seeks speed in the layout of elements to be constructed or installed, yet inaccurate dimensioning may spell disaster if units do not fit together. The designer seeks visual accuracy, structural stability, and quality fit among all components. An owner, as a layperson, most often places trust in the other two parties to see that everything looks and functions as promised.

Ultimate fairness lies in the contract documents—have these documents adequately and clearly shown exactly what is expected of the contractor? Contract documents that fail to state expectations for accuracy are unfair.

The second aspect of fairness is the practicality and adequacy of a designer's expectations. A designer should establish a hierarchy of accuracy to allow flexibility where warranted and precision where necessary. In general, a hierarchy of accuracy will allow flexibility in the measurement of great distances and grow increasingly precise as construction becomes more detailed.

5.3 PREFABRICATION

Prefabricated units include manufactured items specified as supplied and installed by a contractor. Often such items as benches, waste receptacles, exterior luminaires, and prefabricated buildings are purchased and installed by a contractor. A dimensioning plan may inadvertently produce conflict and potential confusion of responsibilities.

A contractor's responsibility will extend only to the proper installation of each prefabricated unit and coordination of its fit with the site. Dimensioning of each prefabricated unit's location and size may become critical in determining ultimate responsibility for fit. Several approaches to dimensioning can be taken by a designer.

5.3.1 Total Contractor Responsibility

A designer dimensionally locates only a prefabricated unit's edge or center and specifies the unit by manufacturer's model number or other reference. The contractor then accepts full responsibility for the unit's exact size and dimensions and constructs whatever size installation framework may be necessary for an exact fit. Detail drawings of the unit do not delineate size. They may indicate the unit's approximate size and installation requirements, but disclaim any accuracy greater than nominal catalog dimensions. Detail or site plans do not indicate the spacing of, for example, footings, bolt ties, posts, vertical depths of footings, or materials. The docu-

ments should clearly indicate that it is the contractor's and manufacturer's responsibility to coordinate detail dimensions of layout, fit, and installation practice. The contractor's and manufacturer's guarantees must also be coordinated by the documents.

5.3.2 Shared Responsibility

A designer fully dimensions and details a unit to be supplied and installed by a contractor. Unless the graphic details carry a provision otherwise, the drawings control all dimensions, materials, and installation of the unit. If the contract drawings differ from the specified manufactured unit, and the unit does not fit, who is responsible? For example, assume that a designer details a prefabricated unit from data supplied by a manufacturer's catalog. Meanwhile the manufacturer changes the spacing or attachment supports and the unit will not fit the space allowed in the site plan. On the one hand, the contractor is required by the technical specifications to supply a specific unit, but the drawings do not reflect the dimensions of that unit. In retrospect, it would have been best to have let the contractor deal with the specifics of the prefabricated unit's dimensions whereas the designer's concern remained with the overall structural and dimensional character and location of the unit.

5.3.3 Confused Responsibility

The three principal parties to a construction contract may become involved in a web of poorly defined responsibilities. In many instances, such confusion results from an overzealous relationship among drawings, specifications, and shop drawings. Shop drawings may be necessary for definition of speciality items requiring extraordinary skill and knowledge with regard to such items as metals, complex water pumping, artwork, and ornamental iron fabrication. It is the designer's responsibility to identify the subjects of a shop drawing. It is the contractor's responsibility to prepare the drawing and specific information for approval by the designer. Most general conditions place the burden of accuracy on the contractor rather than the designer. When a designer calls for a shop drawing, related dimensional information given on contract drawings should be minimized in order to allow a contractor fair responsibility for detailing. It is the contractor's responsibility to check for fit. A designer should not burden the drawings with dimensions that may be impossible to achieve or require a great deal of paperwork to alter. Dimensions should control only the general character and location of the work and allow the contractor full expression of expertise. If the designer completely understood the requirements of the detail, a shop drawing would not have been necessary.

5.4 CONVENTIONAL WISDOM

A dimensional plan is a traditional drawing that has suffered few innovations throughout the years. The drawing is an unembellished presentation of two-dimensional information. Innovation comes hard to a subject rooted firmly in a graphic language that enables communication among surveyors, engineers, landscape architects, architects, contractors, and the various trades. No one may change the language independently of the other's ability to read intent clearly. Convention is necessary for communication. Confusion will reign when someone capriciously and unilaterally introduces techniques unfamiliar to those who must interpret the drawing.

5.4.1 Graphic Clarity

In attaining graphic clarity, the following must be considered.

1. Traditional graphic forms of each numeral have been used through the years to avoid misinterpretation of dimensions. Although a drafter may develop a style of lettering that satisfies the ego and expresses an artistic talent, such individual expressions must be submerged when working drawings are being prepared. The following are examples of numerals that require a conscious and consistent effort during drafting:

   a. 8, 3, 0, and 6 are easily confused;
   b. 7, 4, and 9 are often confused;
   c. 2 and 7 are often confused;
   d. 7 and 9 are always confused.

2. The drafter must distinguish between whole numerals and fractions by using a distinct decimal point. Decimals of a foot are optionally located on the baseline or centered on the numeral's height. Decimals of a meter

are always centered on the numeral's height.

3. The drafter must separate feet and inches with a dash.

4. The use of a foot symbol (') is optional. However, in some situations, confusion will exist between meters and feet. The possibility for confusion between a dash and a decimal point for meters should be noted.

5. The use of a "0" preceding inches or fractions of meters or feet should be avoided. Generally this is an unnecessary effort if decimals and inch symbols are clearly drafted. If an office eliminates symbols for feet, inches, and meters, it may be necessary to precede a fractional numeral with a zero for clarity.

6. Dimensions greater than 1 foot or 1 meter should be given as feet or meters and inches or in decimal form. However, some products are commonly specified in inches of diameter or height. This rule is related to the preceding paragraph.

7. The drafter should avoid touching numerals when underlining, and should leave a clear space between the underline and the numeral.

8. Numerals are to be lettered so that they are legible when read from the sheet's bottom or right-hand side. Multiple sheets are bound on the left edge as a book. Although a left-handed drafter will find this awkward, the convention must be maintained for legibility.

9. Numerals should be written beside, parallel with, and on top of an associated line; for example, with property lines and dimension lines.

10. Fractions are written in only two formats. The choice will depend upon the space available for drafting.

11. One should clearly identify or delineate exactly which point or edge of an object is being dimensioned. An extension line delineates either a point or an edge. When an edge is too small for graphic clarity, it should be noted in words what edge is being referenced. If the dimension is centered, a centerline symbol, a boldly drawn +, or a notation in words should indicate the intended center. A contractor should not have

to interpret to which edge or point reference is being made.

12. Clarity, as well as construction accuracy, is improved if the hierarchy of lines permits the contractor to read and distinguish objects and information. Drawings delineate; they are not simply lines.

5.4.2   Generalities

Some general rules are of interest:

1. Two dimensions are necessary to locate any point in two-dimensional space. All dimensions on drawings must be checked.

2. The dimensioning technique should be matched with the material and function being dimensioned. For example, concrete block and exterior stud walls are dimensioned to their faces whereas piers, interior post positions, and interior walls are dimensioned to their centerlines.

3. Small and delicate lettering of numerals is to be avoided. A horizontal control drawing must be on site where it is subjected to all kinds of weather, soiled hands, and uses that take their toll on legibility and line quality.

4. Dimensions should not appear on both a horizontal control site plan and on details. The site plan drawing becomes cluttered with small dimensions and redundant information.

5. A designer is responsible for delineating all dimensions on the site plan and details. A drafter should not proceed on the assumption that a contractor can go through computations to arrive at the dimensions. Small dimensions should be shown and overall dimensions computed to preclude the necessity for computation in the field.

6. All dimensions should be related to a fixed and locatable starting point. This may seem obvious, however, an area may be so minutely dimensioned that no one can locate it on the site.

7. Local codes and the survey should be checked for fixed and semifixed dimensions that are unchangeable and must control all other dimensions.

8. The system and accuracy of the horizontal control should be related to the methods available or necessary for surveying the prop-

erty. For instance, extreme accuracy requirements, coupled with the need for instrument lines of sight through a forest, may require the cutting of clear lanes through trees for visual sighting.

5.5   RUNNING DIMENSIONS

Running dimensions are common to architecture and detailing of site elements. The system is very simple inasmuch as the drafter is simply identifying a distance between two points by use of extension and dimension lines. Traditionally, the *extension line* is a graphic symbol that extends away from and identifies the object's edge or center. A *dimension line* graphically ties two extension lines together and carries the *numerical dimension*. A dimension line is *terminated* by an arrow or other symbol as it crosses each extension line.

Figure 5.1 portrays the four basic components of running dimensions as a system. Dimensioning begins at the intersection of two fixed property lines. Extension lines are drawn from this beginning point to the object being dimensioned, and a dimension line connects the two extension lines. Each dimension line is terminated as it crosses an extension line. A numerical value in feet or meters is written parallel to each dimension line.

As a system, the dimensions function to locate semifixed buildings from the fixed property lines and then semifixed paving is dimensioned from a building. Note that the running system actually is locating only the corners of all physical objects in Figure 5.1. Because no other information is given in this example, it must be assumed that the building and paving edges run at right angles and parallel to the property lines and that the building dimensions are found elsewhere. The example also indicates a system in which the concrete terrace cannot be located until after the building is located.

Figure 5.1 also serves as an example of graphic legibility as well as symbolism. Note that line weights vary in width, from narrow (light) lines of information to the broader (dark) lines of the physical edges. Numerals are legible and uncrowded. Extension lines bring the dimensions graphically away from the physical objects to avoid crowding the numerals.

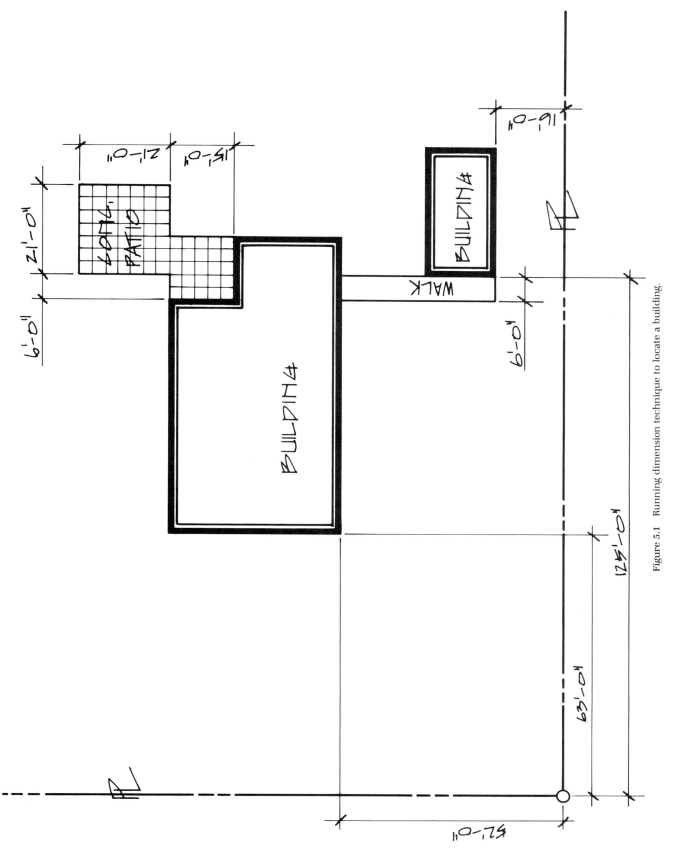

Figure 5.1   Running dimension technique to locate a building.

95

5.5.1 Dimensioning Points in Space

Figure 5.2 illustrates the running system as it functions to locate a point in space. One corner of the building is located. Without the angle of the wall, the bearing of the building's walls could not be found. In this example, one point is found for an object that is not parallel to a property line or other fixed point. A tree is located by two running dimensions that depend upon the building being fixed in space. The light post can be located at any time during construction because it, like the building corner, is dimensioned to fixed property lines and is independent of the building dimensions or location.

5.5.2 The Extension Line

Following are a few conventions regarding extension lines.

> Extension lines are drawn very sharply and about $\frac{1}{16}$ inch short of touching the object or point being dimensioned.
>
> Extension lines may cross only other extension lines.
>
> Extension lines may be drawn through the object being dimensioned, if this increases clarity and is less ambiguous than trying visually to trace the extension line's location.
>
> A centerline or property line may serve as an extension line.
>
> Extension lines are continuous and are not broken at a terminus with a dimension line.
>
> Extension lines should be used to locate dimensions outside of the object being dimensioned.

Figure 5.3 illustrates a few examples of the use of an extension line combined with dimension lines. The widths of the wall and fountain wall are not given because these dimensions would ordinarily appear on a detail drawing. For example, if the detail specified concrete block, the wall's width would be automatically dimensioned by the block's nominal dimension. Note that the location of extension lines has been planned to avoid the crossing of dimension lines and that extension lines are graphically located outside of each object for clarity. Extension lines are taken consistently from the face of the wall to the object.

5.5.3 The Dimension Line

A few conventions regarding dimension lines are as follows:

> Dimension lines are drawn very sharply and either to or across an extension line.
>
> Dimension lines never cross one another. When dimension lines are crossed, it is often symptomatic of poor graphic organization. If dimension lines must cross, one of the lines should be broken or bridged so that the crossing is obvious (see Figure 5.4).
>
> All dimension lines are terminated at an extension line. Several graphic styles have been developed, such as Figures 5.5 and 5.6. The use of arrows requires more precision than many drafters are willing to give. The alternatives function as termination points or slashes and require a line weight several times darker or thicker than the extension line. Unless a drafter actually crosses an extension line and dimension line, these alternative terminations will not read well and take more time to draw than arrows do. Highly stylized terminations should be avoided; these, in great numbers, confuse the drawing. When ink is used, alternative termination symbols will require two different pen widths.

Dimension lines should be spaced $\frac{3}{8}$ inch apart, depending on the height of the numerals, and a consistent spacing maintained between dimension lines throughout the drawings.

Numerals are conventionally located on *top* of each dimension line, which makes them easier to find.

When dimensioning a cut detail drawing, the dimension line should be broken or drawn as a wave. See Figure 5.8 for an example.

5.5.4 Details and the Site Plan

Figures 5.7 and 5.8 illustrate graphic coordination among site plans and details. Figure 5.7 is an example plan view of a wooden deck and overhead structure. Note that dimensions are given to the centerline points of each pier upon which the overhead structure and a portion of the deck will rest. The building's corner serves as a fixed reference point from which measurements will be taken and

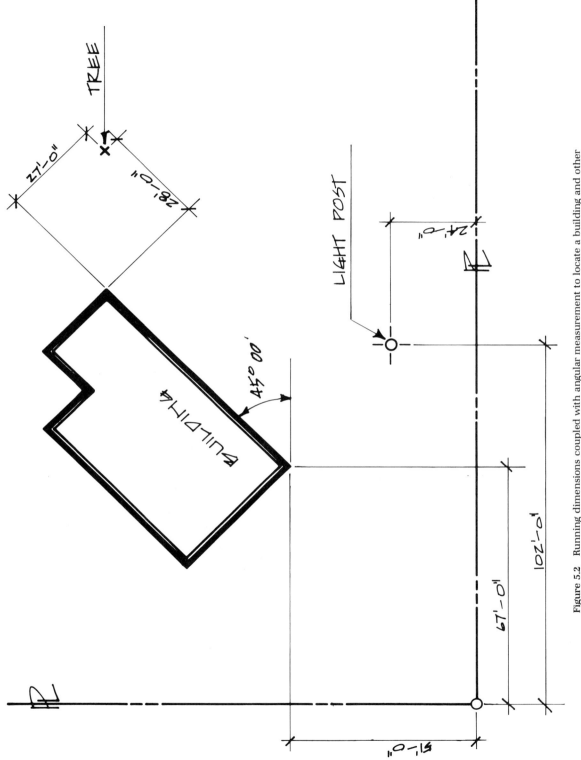

TREE

27'-0"

28'-0"

BUILDING

45° 00'

LIGHT POST

24'-0"

102'-0"

67'-0"

51'-0"

Figure 5.2 Running dimensions coupled with angular measurement to locate a building and other construction.

97

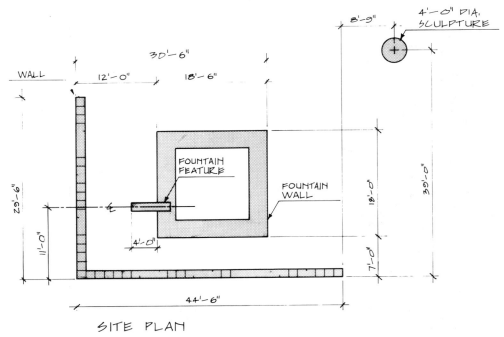

SITE PLAN

Figure 5.3   Running dimensions used to locate construction items.

to locate the structure on the site. With reference to Figure 5.8, note that the post pier dimensions are given from a centerline comparable to the plan's center point. Once a contractor locates the center points, other dimensions are found from the detail.

trigonometry. Historically, these systems involved time-consuming computations and were useful only for construction requiring very precise locations. However, modern calculating equipment makes computation quite easy, and very useful for almost any site's horizontal control.

5.6   LINEAR DIMENSIONS

One group of techniques involves knowledge of an imaginary line's direction (bearing) and the location of a point along such a line. These are the only two dimensions necessary for on-site location of a point.

An imaginary line may be either straight or curved. Straight and curved lines are found and related by

5.6.1   Baseline System

A baseline system may be both imaginary and real. For an imaginary line, a surveyor visually orients a line of sight and measures points along that line. For a physical line, a contractor might stake two points and stretch a string between them as a physical line of reference. Either way, a drafter is simply present-

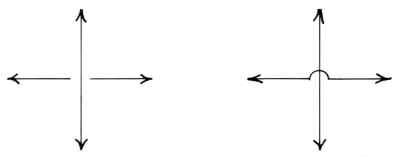

Figure 5.4   Two techniques useful in avoiding crossing of dimension lines.

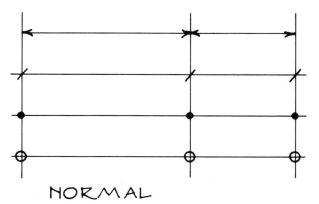

NORMAL

Figure 5.5 Four acceptable symbols to terminate dimension lines at extension lines. The most traditional and acceptable symbols appear at the top and the least acceptable at the bottom.

ing information that can be measured along a baseline. A baseline can be either a fixed dimension or a semifixed dimension, depending upon the graphic presentation and the precision used to find it. Unless stated otherwise on a drawing, a contractor will always assume lines to be at right angles to each other.

Figure 5.9 illustrates a plan view of a typical site improvement that is dimensioned in baseline technique. The drafter assumes that a horizontal bench mark may be located as a beginning point. A baseline is fixed at a right angle or specific bearing to the bench mark and running dimensions locate the various elements along the baseline. A baseline such as this can extend to infinity and works quite well for linear-shaped sites.

Figure 5.10 illustrates the baseline system as it might be used to find irregularly spaced points in space. Connecting these points in space will pro-

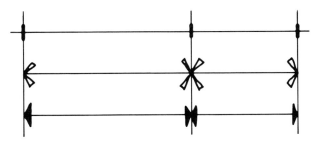

Figure 5.6 Three termination symbols that are unacceptable because they often confuse, overdraw, or consume drafting time.

duce a facsimile of the desired physical shape. Construction accuracy is limited mainly by the number of points found to describe the physical shape. Graphic accuracy is limited only by the drawing's scale and space in which to write information. Note that each point is made graphically visible by being drawn darker than the perimeter shape of the physical object to be constructed. In normal use, it is assumed that a designer obtains dimensional data by scaling a drawing, if the physical shape is nonstructural and of arbitrary design. A contractor is usually allowed some latitude in field adjustment. A designer should note the degree of accuracy necessary during layout and any designer approvals or adjustment that might be required prior to construction. Overall dimensions are not ordinarily given because each point is located independently of the others, unless the site's dimensions constitute a restriction.

Figure 5.11 illustrates a baseline system coupled with a stationing system. (See Figure 5.14 for an explanation of the stationing system.) Note that the number of running dimensions has been reduced considerably. Running dimensions are used only for those distances that are offset at right angles to the baseline. Distances running with the baseline are stationed. The drawing will be less cluttered when dimensions and graphic conflicts are reduced to a minimum. Note, however, that the stations require contractor computation of distances between any two points running with the baseline.

5.6.2 An Offset System

An offset system differs from the ordinary baseline system only in that each dimension running parallel to the baseline is found at a regular interval. A designer must bear in mind that the location of points in space has only a chance relationship to the dimensioned shape, that is, the opportunity to select and locate important or critical points in space does not occur. Because of these factors, the final physical construction cannot be expected to be precise.

Figure 5.12 illustrates a simple offset system. Dimensions are scaled from an arbitrary and nonstructural shape. Accuracy is dependent upon the number of dimensions given, the shape's complexity, and the drafter's skill in drawing and scaling. The system's major advantage is speed in both drafting and on-site layout.

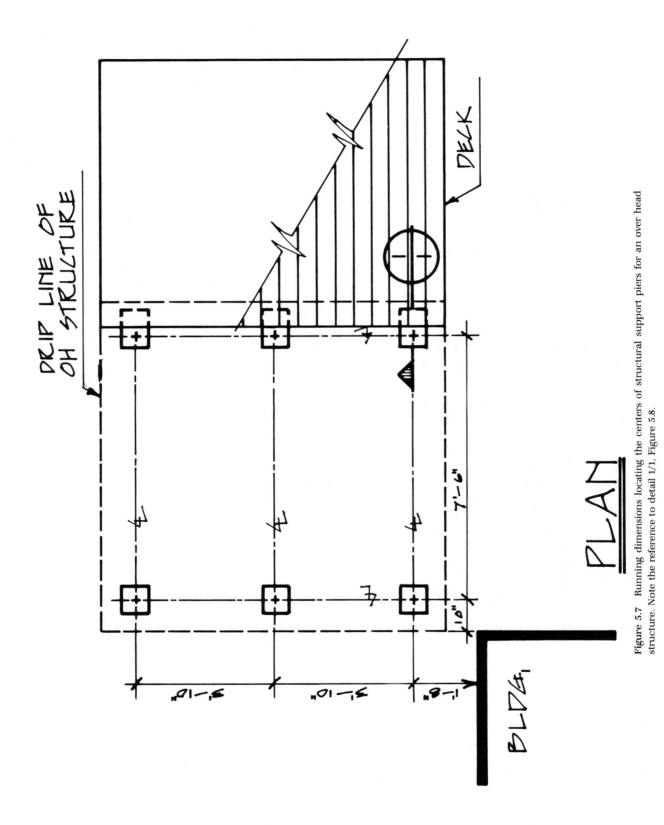

Figure 5.7 Running dimensions locating the centers of structural support piers for an over head structure. Note the reference to detail 1/1, Figure 5.8.

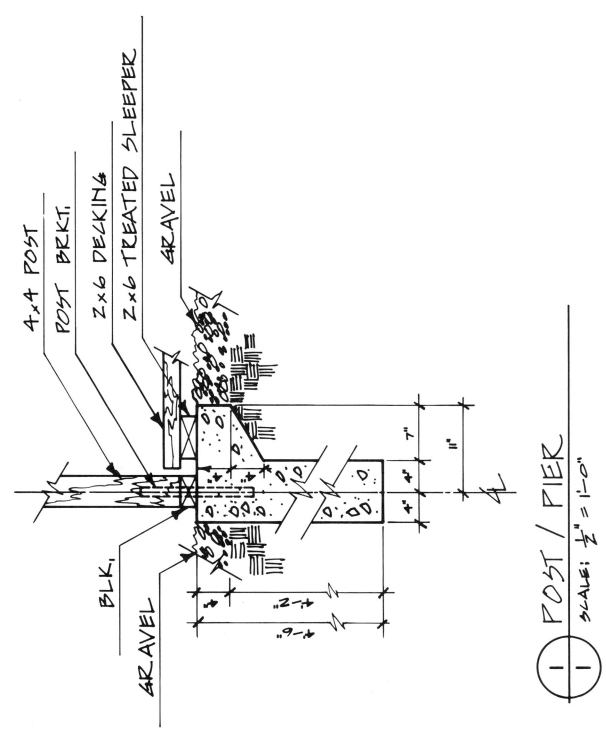

4x4 POST

POST BRKT.

2x6 DECKING

2x6 TREATED SLEEPER

GRAVEL

BLK.

GRAVEL

7"

11"

4"

4"

4'-2"

4'-6"

POST / PIER

SCALE: $\frac{1}{2}$" = 1'-0"

Figure 5.8   A sectional detail of the pier supports for an overhead structure noted in Figure 5.7. Note that the detail dimensions complement the centerline location of Figure 5.7, that dimension lines complement the drafter's break in the pier, and the variations in line weight.

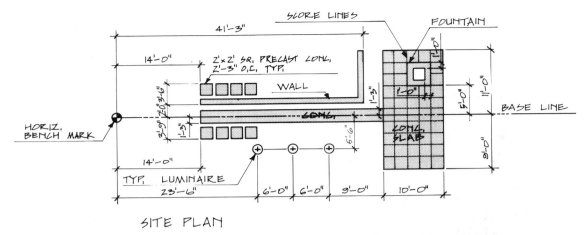

SITE PLAN

Figure 5.9 The use of running dimensions from a baseline. Note the use of the symbolic *o.c.* as a dimensioning technique whenever units are typically (typ) evenly spaced, and the technique of drafting offset dimensional numerals whenever space is too limited for normal numeral location. The example presumes that a benchmark is located from some known point(s). Dimensions, such as a wall's thickness, remain for large-scale details.

5.6.3   A Grid System

There is a natural progression in moving from a single baseline to two baselines. When two baselines are delineated at right angles to each other, a grid becomes the next step in the process.

A grid simply provides a framework within which a complex shape can be developed without straying too far from the contemplated shape. Although the system appears to be "loose," it may be the only means to establish the scope of construction and cost and yet maintain a degree of flexibility in laying out a complex and irregular shape. Figure 5.13 illustrates a grid system using a 10 × 10-foot grid. If a drafter required increased accuracy, each grid could be made smaller or any one of the grids could be subdivided into smaller grids.

5.6.4   Stationing

A stationing system is based upon graphic and survey techniques that identify and locate distances along an imaginary baseline. Such a line is often used to lay out streets and highways, can be straight or curved, and is useful in site work when combined with other systems (see Figures 5.11 and 5.33). Graphic clarity and use of conventional surveying techniques are among the system's advantages; a potential accumulation of error can be a disadvantage.

Figure 5.14 illustrates a stationing system's components. A baseline, the direction (bearing) of the baseline, a dimensional value for a specific station point, and regular index stations along the baseline compose the major graphic information. Note that stations A, B, and C are potential locations for some item to be constructed or otherwise located. Numerals are traditionally written as, for example, 1+ 33.25', which is read as 133.25 feet. This graphic form replaces the comma as a means of separating each 100- or 1000-foot unit interval thus helping to avoid major computation error. Index stations are even whole numbers graphically located by scale along the baseline's length. The index number aids in visually identifying an error in a station's location or numeric value. For example, if a mathematical error were to compute station B as 3+33.25, it would be obvious that an error had occurred because point B must be a numeric value with a magnitude greater than 1+00 but less than 2+00. Such an error should be identified at the time of drafting. This is a good reason to plot index numbers on the baseline before stationing specific points in space.

A baseline may be given direction by the use of an angle, as Figure 5.14, or in the form of a bearing. (See Section 5.6.5 for a discussion of bearings.) Figure 5.14 illustrates running dimensions used to locate the beginning point of a baseline. This point is noted as 0 + 00 and bears to infinity 105 degrees off the wall of an existing building. All subsequent stations along the baseline are computed or scaled from 0 + 00 and are accumulative.

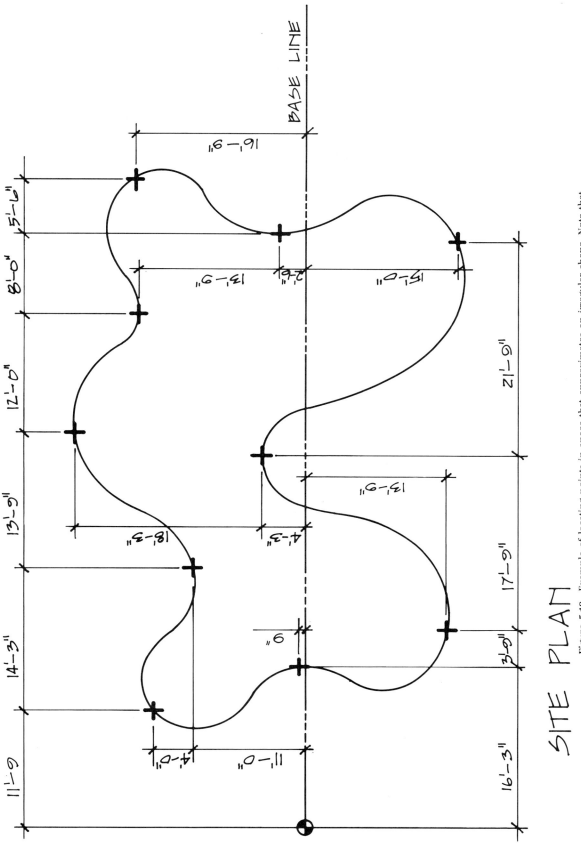

SITE PLAN

Figure 5.10 Example of locating points in space that approximates an irregular shape. Note that running dimensions are generated from a benchmark and baseline. Line weight variation is necessary to maximize legibility of the drawing. Points are located only at the apex of concave and convex curves as guides to layout; more points will increase the accuracy of layout.

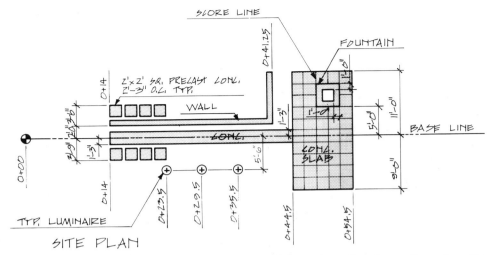

Figure 5.11 Geometric dimensioning technique using both running and stationary dimensions. This example is similar to Figure 5.9, with stations replacing all longitudinal running dimensions in order to simplify the drawing.

Figure 5.15 "reads" as follows: Beginning at station point 1+09.32 on the centerline of the existing street, turn an angle of 80° 00′00″, measure 98.00 feet on that line, and turn an included angle of 110° 00″ for a distance of 127.00 feet (2+25.00 less 0+98.00′) to locate the end of the drive's construction. Turn an angle of 25° 00′ and measure off 20′ − 00″ along that line. Turn a 90-degree angle and measure 25′ − 00″ to locate the corner of the feature. The edge plane of the feature lies at 25° 00′ relative to the entry road Figure 5.15 illustrates the finding and layout of a drive and relates the location of a construction feature to that road. The accuracy of such a method depends upon the accuracy of computing and locating the angles and lengths of lines.

Because stations are computed and found on horizontal dimensions, they can be used in either a plan or profile view. For example, the stations indicated on Figure 5.15 would be the same whether shown as a plan or a profile.

Figure 5.16 illustrates a graphic profile of a road, walk, or any other linear baseline that is "stretched out" in profile view. Note that several items, such as a fireplug and light fixture, are located on the profile. In essence, these items are dimensioned almost as fully as though they were shown in ordinary plan view. A drafter can use profile or plan for locations, except that items located on profile must indicate whether they are to be offset and in what direction they are offset from the baseline. If items are consistently offset from a baseline, such information may be indicated on a detail or schedule. Again, the drafter must avoid redundancy or conflict between the plan and profile views.

5.6.5 Bearings and Angles

The bearing of a line is its horizontal location with respect to true north. Conventional usage expresses a bearing as, for example, N 32° E and translates to mean a line that bears in an easterly direction 32 degrees from true north. Conversely, a bearing of S32° W would mean the same line, except bearing in a westerly direction from south. The value of a bearing, in degrees, is then relative to some point of beginning. Property lines are described by bearings in a circular fashion (a traverse), that is, beginning at one point and following bearing data in one direction until returning and closing upon the beginning point. Figure 5.17 illustrates the relative language of several lines bearing in relation to true north.

Bearings are critical to the relationship of fixed dimensions and as an alternative means of indicating an angle between two lines or objects. Bearings should be used whenever the construction specifications require a registered land surveyor to execute fixed lines or lay out the construction work. Bearings are the most common language of surveying and a full discussion of their use and procedures can be found in most surveying tests.

Figures 5.18 and 5.19 illustrate two graphic means and symbols for indicating the bearing of a line.

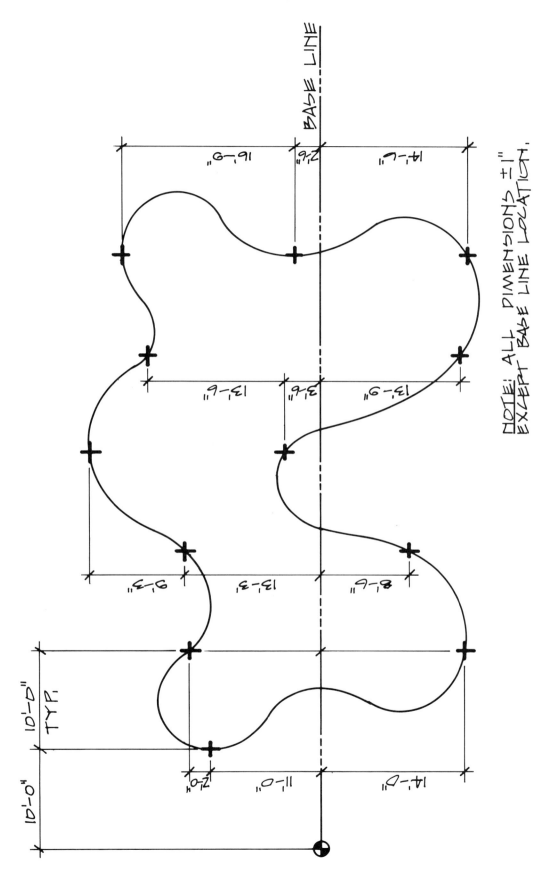

Figure 5.12 Irregular shape dimensioned by spacing points in space at regular (typical) 10-foot, 1-inch intervals along the baseline. Note the statement of acceptable accuracy. Increasing the number of points in space will increase the accuracy of the layout.

SITE PLAN

105

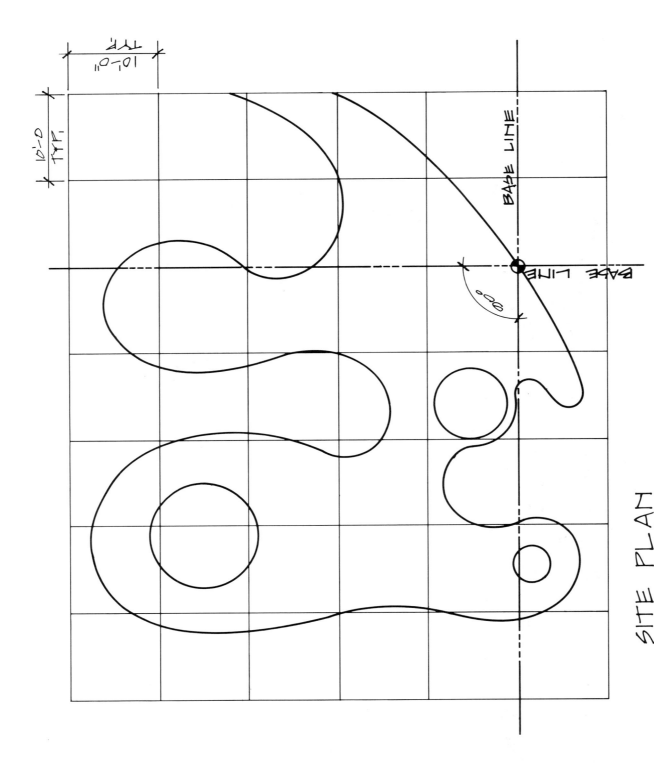

10'-0"
TYP.

10'-0"
TYP.

BASE LINE

BASE LINE

90°

SITE PLAN

Figure 5.13 Dimensioning of a complex irregular shape by formation of a regular grid with typical (typ) dimensions. The benchmark would be located from a known point(s) and the grid established on site. Once the grid is physically laid out, the irregular shape will be estimated within each grid. This is useful only when precise measurements are not necessary.

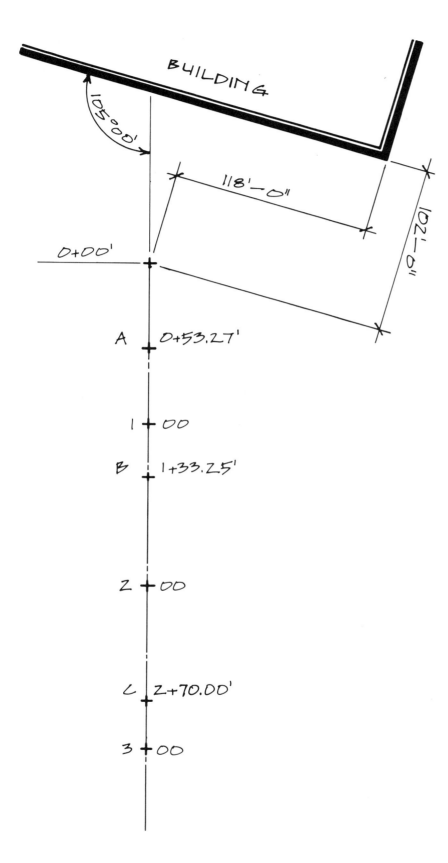

Figure 5.14 Development of a baseline at a particular angle to a base object. Note that points are located by stationing along the baseline's length.

Figure 5.18 illustrates the conventional manner of presenting the value of a bearing and Figure 5.19 shows the use of a standard drafting technique with angles. Both presentations indicate the same information but in a different manner.

Figure 5.20 illustrates a graphic situation that may often make the job of surveying more difficult than it need be. In delineating the angle as Figure 5.20, it is impossible physically to set up an instrument at the corner of an existing vertical wall or element in order to turn the 130-degree angle. Although it may appear feasible on paper, a surveyor will be forced to use other ways of finding and locating the baseline. New work may be located directly from such graphic information because there is no physical deterrent to instrument location. The use of latitude and departures can be very helpful to the surveyor, contractor, and drafter faced with a situation similar to that shown in Figure 5.20.

5.6.6   Latitudes and Departures

Computation of latitudes and departures allows angular descriptions to be simplified into easily accomplished running dimensions. Quite often, it is impossible for the contractor's surveying instrument to turn angles with a high resolution of accuracy. A designer can compute running dimensions for use in finding precise angular measurements. In principle, a line that has a bearing and

length is viewed as the hypotenuse of a right triangle. The *latitude* is the computed height of a triangle bearing as north. The *departure* is the other leg of the triangle and bears east. The equations for computations are

$$\text{departure} = \text{length of hypotenuse} \times \sin \text{bearing angle}$$
$$\text{latitude} = \text{length of hypotenuse} \times \cos \text{bearing angle}$$

In more precise terms, the departure of a line is its length relative to east and latitude is its length relative to north. For example, a baseline bearing N 70° 00′ E and 150 feet in length has a latitude (north) length of 51′ 3⅝″, a departure (east) of 140′ 11½″, and can be dimensioned as Figure 5.21.

A contract drawing should carry one type of information—the bearing and length of the baseline *or* the running dimensions of latitude and departure.

Figure 5.22 demonstrates one use of latitude and departure information as applied to the land surveyor's problem illustrated in Figure 5.20. The baseline can be located from the building's corner by the drafter's use of latitude and departure information. Running dimensions are computed and drafted for the contractor's use in finding both the bearing and the beginning of the baseline. The length of 100 feet is arbitrary and selected only to give tape measure accuracy to the baseline's bear-

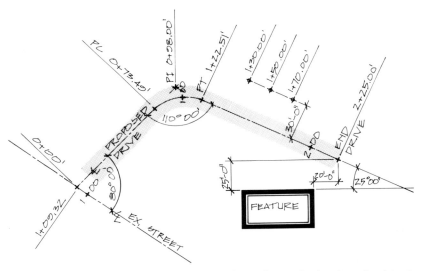

Figure 5.15   Demonstration of the use of angles to locate the bearing of a driver's centerline. Note the use of stationing distances computed to locate PC/PT and end of work, the stations that locate items along the centerline, and that a "feature" is located on site by angles and running dimensions.

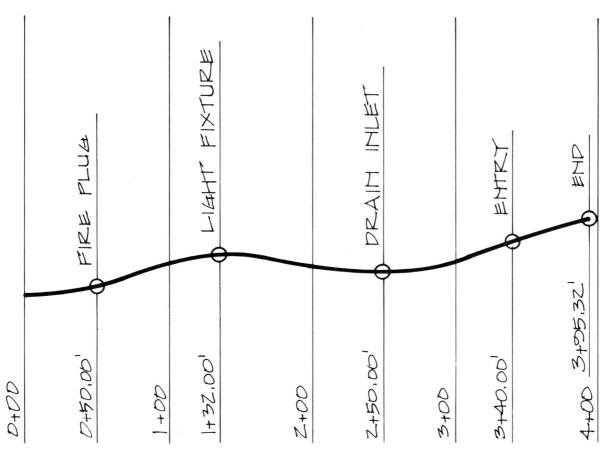

Figure 5.16 Simplified longitudinal profile of a vertically curved road or walk indicating a technique for recording horizontal distances along the profile. Note that various site features are also located by station. A plan view would be necessary to indicate the offset distance for each feature along the profile's line.

ing. Also, the angle of computation in this instance is 50 degrees (180 − 130 = 50 degrees). The bearing of the baseline can be found and checked by the use of running dimensions and a tape measure. The angle of 40 degrees given in Figure 5.22 is not necessary on the contract drawing, although it may be useful as a double check in backsighting an instrument along the baseline.

Figure 5.23 illustrates the use of running dimensions to locate a building or paving area on a site with specific setbacks or other reasons for its precise location. Again, running dimensions allow layout without dependence on an instrument for the turning of bearings. It is the designer's intent to position the 40 × 80-foot paving at a bearing of N 52° 00' E. Note that the bearing need not appear on the drawing. Also, the dimension of 40 × 80 feet is not precise in this example because of rounding off of latitude and departure data. The contractor is

notified of this accuracy by a plus or minus paving dimension.

5.6.7 Curve Data

So much has been written regarding the use and computation of curve data that no attempt will be made here to cover the subject comprehensively. However, the manner in which curve data are presented and the technique's application to horizontal control situations is of interest.

In principle, the use of curve data is a variation on baseline, baseline bearing, and stationing. In application, the technique allows a line to change direction, with precise location of such a change. Common convention refers to the straight baseline as a *tangent*. Tangents, like a baseline, are given direction by a bearing, and when changing direction, by a deflection angle. A tangent's length is given value

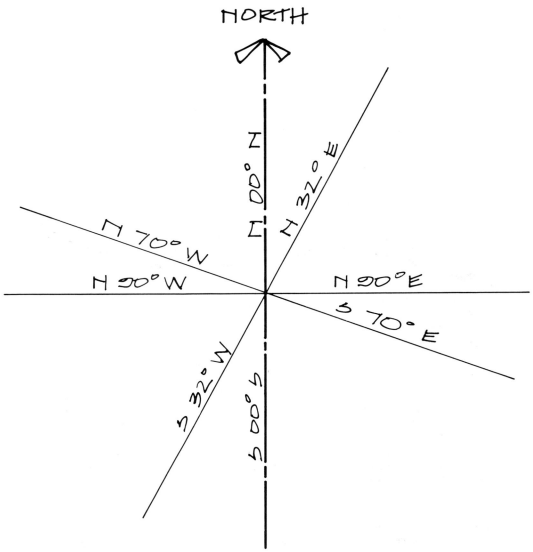

NORTH

N 00° E

N 32° E

N 70° W

N 90° W

N 90° E

S 70° E

S 32° W

S 00° E

Figure 5.17   Several examples of the language and direction of angles stated in the form of "bearings."

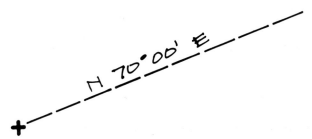

N 70°00' E

Figure 5.18   Graphic plan form of a bearing indicating the direction of a dashed line from point +. Note that a bearing provides only direction and not length of a line.

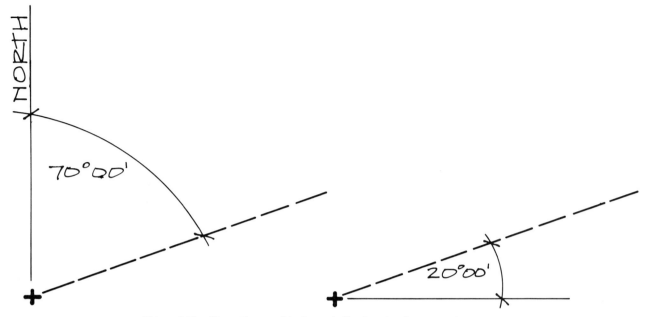

Figure 5.19 Alternative graphic forms indicating the direction of a line.

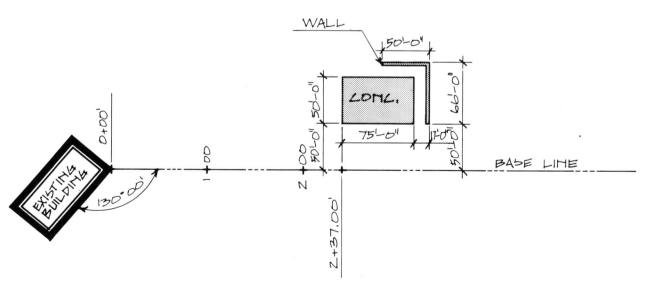

Figure 5.20 Relationship of several site features by dimensioning from an established baseline and incorporating angles, running dimensions, and stations.

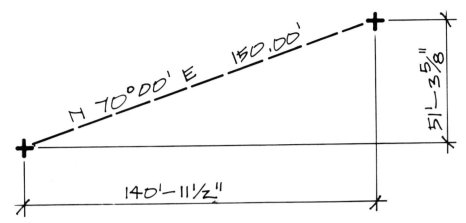

Figure 5.21 Relationship of latitude and departure data computed from a line's bearing and length. Normally a document will indicate either length and bearing *or* latitude and departure data.

through the use of stations. Stationing values are relative to the length of the curve. A curve's length depends on its radius and deflection angle.

Figure 5.24 indicates the basic types and graphic location of information required for a horizontal curve layout. Each of the PC and PT stations requires a numeric value regarding its respective distance from 0+00. A deflection angle is necessary for the bearing of each tangent. The length of each curve from the PC and PT points is also necessary. To find the station PT of curve 2, a drafter adds the lengths of the tangent between 0 + 00 and station PC of curve 1, the length of curve 1, the length of the tangent between station PT of curve 1 and station PC of curve 2, and the length of curve 2. Each point of tangency is given a station number by adding only the tangent distances directly between each point of tangency.

The graphic information is abbreviated and suitable for a surveyor's interpretation. Certain aspects of the graphic format will vary from office to office, but Figure 5.25 is a typical example. Note that curve data are placed in a box or otherwise clearly separated from other information. On occasion, curve data are placed in a schedule, with each curve identified by letter or number. Information is given in a straightforward manner with no graphic embellishment. Tangent point stations may also be added to the drawing. The bearing of each tangent may be given. The drafter should view the drawing as simply a means of organizing information. A surveyor does not depend upon the drawing because all layout information will be taken from computed data. The line illustrated by Figure 5.25 could represent the centerline or edge of a road, drive, walk, grassed area, or any other site feature that is linear in nature.

5.7 GRID COORDINATE SYSTEM

The grid coordinate system incorporates certain aspects of all previously discussed horizontal control systems. The grid itself is composed of two right-angle baselines; running dimensions or latitude and departure can be used to locate elements in the grids.

Figure 5.26 illustrates the basic concept of a grid coordinate system. Commonly the intersecting grid lines run north and east with a beginning or bench mark located in the lower left-hand corner of the site. Such a grid relates the trigonometric *y*-axis to latitude and *x*-axis to departure, and maintains computations in the positive quadrant with respect to the *xy*-axis.

Although it is common practice to maintain the grid in relation to a specific site for ease of layout, it is possible that some sites will relate to the larger context of a U.S. Coast and Geodesic Survey grid network or other regional or state grid systems, such as local property section lines. Large sites with partial ongoing developments may require the location and construction of on-site monuments for a permanent recording of grid intersections and ongoing surveying of each site work contract. These special needs should be made a part of the owner's information or a contractor's portion of the construction horizontal control.

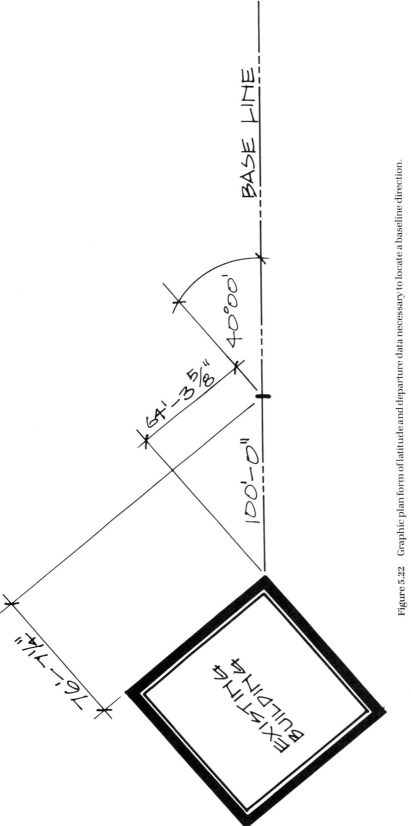

Figure 5.22   Graphic plan form of latitude and departure data necessary to locate a baseline direction.

113

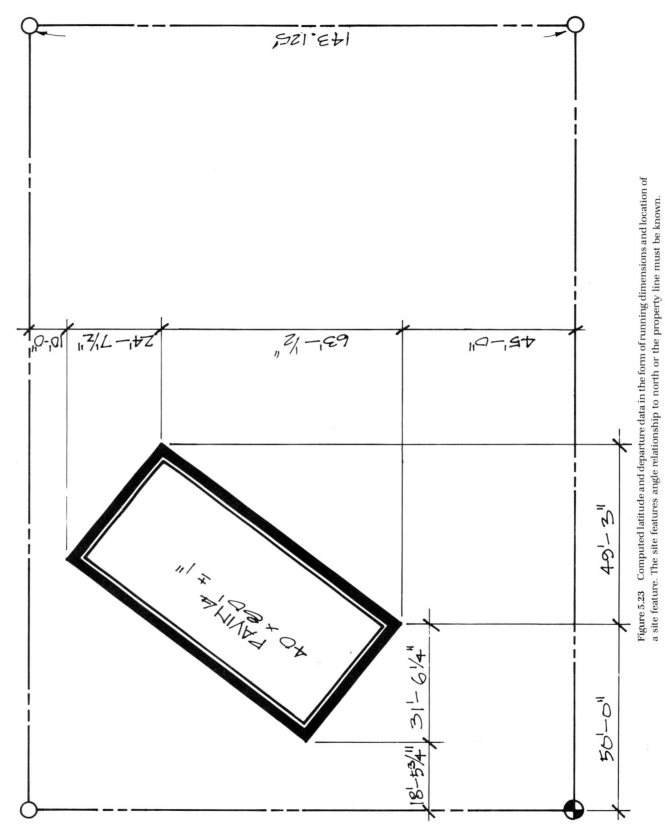

Figure 5.23 Computed latitude and departure data in the form of running dimensions and location of a site feature. The site features angle relationship to north or the property line must be known.

114

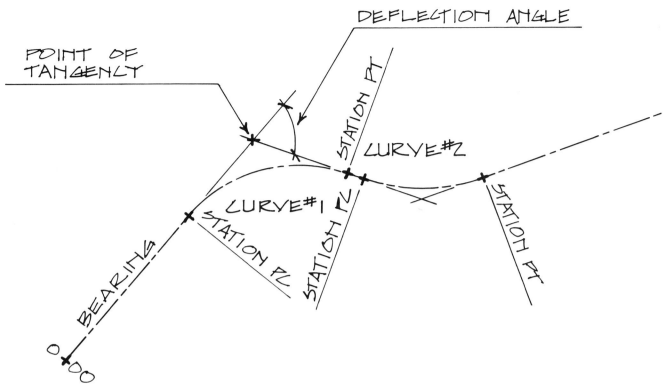

Figure 5.24 Horizontal curvature of a road or walk in plan view. Note the fundamental graphic data and data location when drafting similar conditions. The deflection angles could be given in angular or in bearing form.

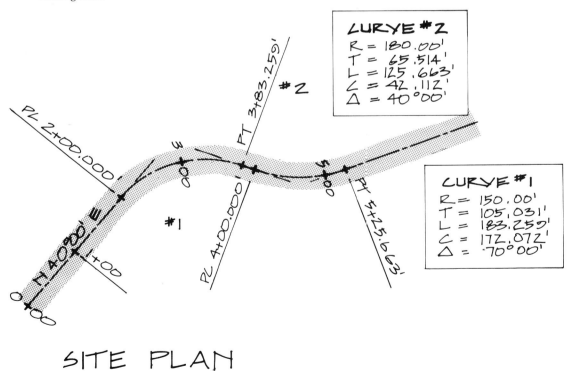

CURVE #2
R = 180.00'
T = 65.514'
L = 125.663'
C = 42.112'
Δ = 40°00'

CURVE #1
R = 150.00'
T = 105.031'
L = 183.259'
C = 172.072'
Δ = 70°00'

SITE PLAN

Figure 5.25 Example of a curved road in plan view indicating proper graphic form of data. Bearings of each tangent line could be added in bearing form.

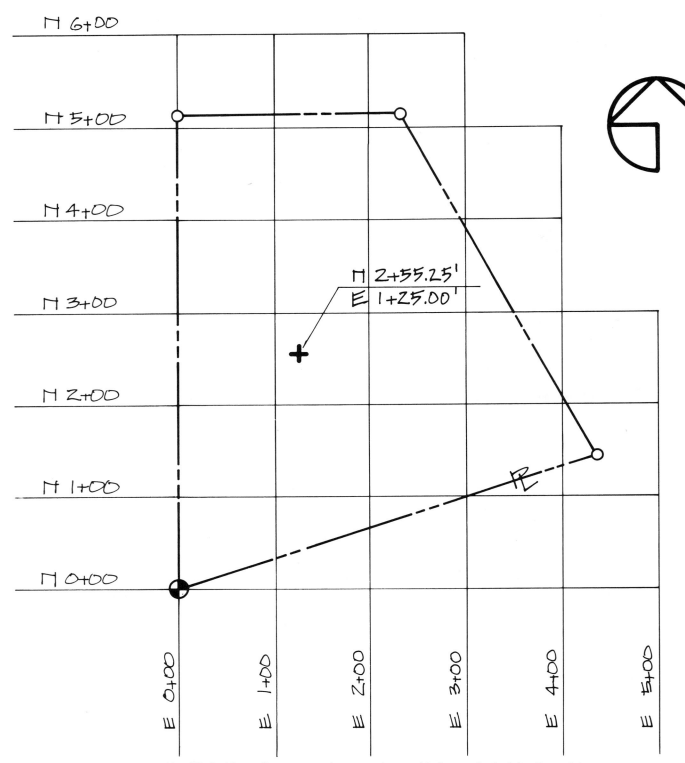

Figure 5.26 Simplified grid coordinate system demonstrating graphic form and principles. One point in horizontal space is located by northings and eastings.

116

Figure 5.26 specifically locates a single point in space. Note that the point is given in much the same numeric form as stationing. The letter N relates to those grid lines running east and west. However, note also that the letter N corresponds to a distance north from the bench mark, and that all grid lines bear a distance north from the bench mark. The point is then read as being 255.25 feet due north of the bench mark, lying between grid lines N2 + 00 and N3 + 00. The other grid lines are similar in that they run north and south and identify a distance to the east of the bench mark. The number E1 + 25.00′ is read as being 125.00 feet due east of the bench mark and lying between grid lines E1 + 00 and E2 + 00. The bench mark is designated as 0 + 00 in both the north and east directions.

It is interesting to note that the grid coordinate system is both a graphic and verbal source of information. Two people may communicate only one point in space without fear of a mixup in direction or distance. For example, if one of the property corners is located at E4 + 34.94 feet and N1 + 49.76 feet, such a location may be directed or described verbally, graphically, or in writing.

Several conventions exist in drafting and terminology. A grid is usually drawn very lightly as a halftone background. The grid serves only as a visual means of checking one's work in the same manner as index stations on a centerline. Each grid line is identified as to its horizontal distance from the bench mark. Dimensions that run from the bench mark north are commonly referred to as *northings* and those running east as *eastings*. The graphic presentation of northings and eastings should be as shown on Figure 5.26, with the exact point of reference indicated by a dark cross mark, dot, or arrow. If a large number of northings and eastings is graphically necessary, it is advisable to clearly enclose each pair in a box.

When proposed construction lies at a defined angle to the grid, data must be computed as latitudes and departures. Figure 5.27 demonstrates the use of grid coordinates placed at a site feature's corners. The feature is 40 × 80 feet and translates the latitude and departure data of Figure 5.23 into grid coordinate language. Note that subtraction or addition of the latitude and departure data of Figure 5.23 is used to arrive at the grid coordinates. Only a scaled relationship exists between the grid itself and the grid coordinates. However, the grid serves as a

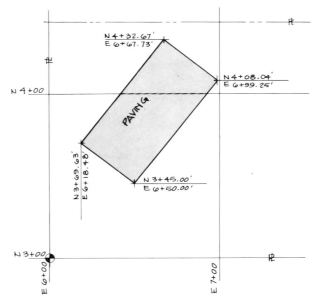

Figure 5.27 Graphic and data demonstration locating a site feature within a grid and using coordinates. A designer can, for example, arbitrarily locate corner N 3 + 45.00′/E 6 + 50.00′ and then determine the remaining three corners by computing latitude and departure distances. Latitudes and departures are translated into coordinate form. The site feature's bearing to north and its dimensions must be known.

visual check of the computations and the feature's relative location.

Figure 5.28 demonstrates the grid coordinate system to locate one corner of a site feature. The corner of the feature is located, a bearing or angle given to one side, and the feature's dimensions given by a running or nominal dimensional system.

Adherence to certain principles is necessary to avoid conflicting and incorrect information appearing on a drawing. For example, the initial coordinate location of a feature's corner, as in Figure 5.27, can be arbitrary in its location but, once set, the three other corners must be determined by computation and cannot be set by scaling. Attempting to scale latitudes and departures would appear to be correct as coordinate data but would probably produce a construction feature with dimensions other than 40 × 80 feet.

If latitudes and departures are scaled from a drawing, the inaccuracy of scaling will produce points that are not in a straight line. For example, if a row of luminaires were to be located along a line

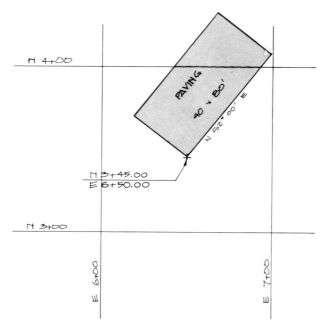

Figure 5.28 An alternate data presentation useful for locating a site feature within a grid system. The coordinate is arbitrarily determined, the feature's dimensions given, and one side of the feature given an angle to the grid.

bearing N 80° 00″ E and they were spaced 20.00 feet apart, each computed latitude would be 3.47 feet and departures would be 19.69 feet. A construction drawing of the locations might appear as Figure 5.29. Obviously it is impossible to obtain such accuracy with a scale when each feature must be located 20 feet on center.

Figure 5.30 illustrates a grid coordinate system adapted for use in locating arbitrary or other scaled dimensions for reproducing irregular shapes. As long as curves do not close with any geometric precision, the coordinate points may be scaled as to location. Note that the coordinates are simply an alternative system to that illustrated by Figure 5.12.

Figure 5.31 illustrates an unusual but sometimes useful variation of the baseline and coordinate systems. A framework of selected lines of sight and turning points is designed in the manner of a coordinate system, except that the northings and eastings are related to spatial visual openings in vegetation or topography rather than an arbitrary geometric grid. Whenever possible, each turning point is located as the center of a green or tee or other landscape feature. Once a framework of turn-

ing points is established, the distance and bearing between them may serve as a baseline from which other landscape features may be located. Turning points may be identified by either temporary or permanent markers for use during and after construction. Note that in Figure 5.31 coordinate dimensions are given in whole-number feet, implying an acceptable accuracy of ±0.5 foot for all measurements in the example.

5.8   COMBINATIONS OF SYSTEMS

Figures 5.32, 5.33, and 5.34 demonstrate three different approaches to the same horizontal control problem. Each example serves either as a relatively faster, more accurate, or more complete method of drafting and presentation.

Figure 5.32 is standard procedure but suffers slightly from the number of dimensions required and the difficulty in achieving the road's curve layout. The PT line is assumed to be a center of the 19-foot-wide parking bay, but the drawing is not explicit. No curve data are given. Also, radii for curbs within the parking area are not included. One confusing piece of information is the 7′ − 11″± dimension next to the drive. The lot is 170.4 feet wide and adding the parking and road dimensions produces 170.5 feet. It must be assumed that the contractor is to determine where to reduce the collective dimensions by 0.1 foot. A contractor is committed to begin the layout on the lower property line because a survey may not begin on a plus or minus dimension. Some dimension lines cross unnecessarily.

Figure 5.33 combines stationing, running dimensions, and curve data. Parking bays and islands are located by centerlines and stations. Crossing dimension lines is avoided. Curve data fix the road's curve. Island curb radii are indicated by notations. Fewer dimensions are necessary than in Figure 5.32, but better overall accuracy and information are achieved.

Figure 5.34 is, again, a variation. Crossing of dimension lines is eliminated by the use of stationing centerlines. Several running dimensions are eliminated by stationing. Curve data are given. A general graphic simplicity is achieved, with an increase in accuracy and information when compared with Figure 5.32.

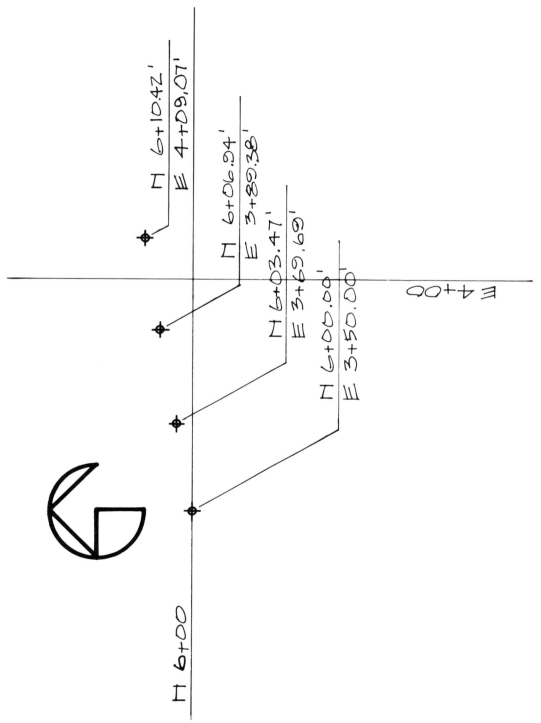

Figure 5.29 Computed coordinate data given to four objects that must be installed in a straight line and related to a grid. *All* data must be computed, except that one point may be given an initial arbitrary location by coordinate.

119

120

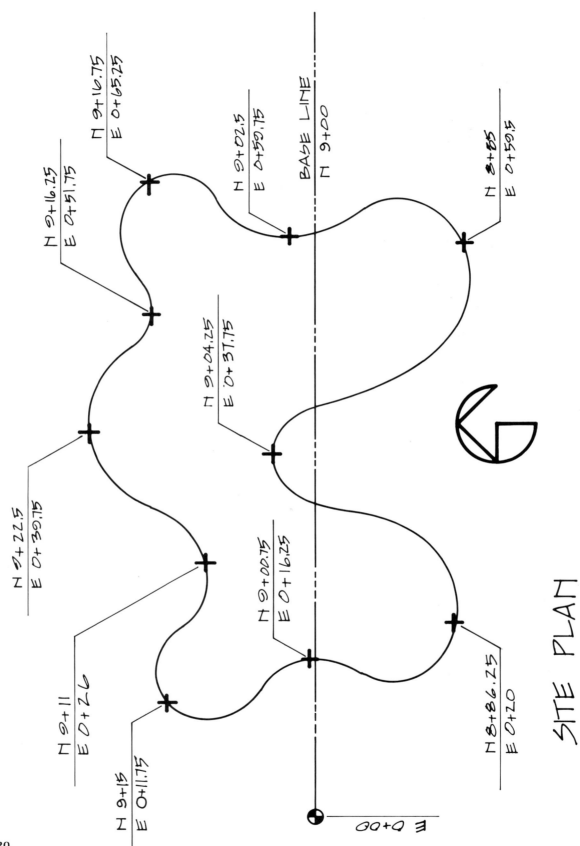

N 9+16.75
E 0+65.25

N 9+16.25
E 0+51.75

N 9+02.5
E 0+59.75

BASE LINE
N 9+00

N 8+85
E 0+59.5

N 9+04.25
E 0+37.75

N 9+22.5
E 0+39.75

N 9+11
E 0+26

N 9+00.75
E 0+16.25

N 9+15
E 0+11.75

N 8+86.25
E 0+20

E 0+00

SITE PLAN

Figure 5.30   Graphic form of dimensioning an irregular shape by using coordinate data. Unless the curves are generated from precise data, all points in space may be scaled as to approximate locations. The baseline may be one grid line or located with respect to the grid system.

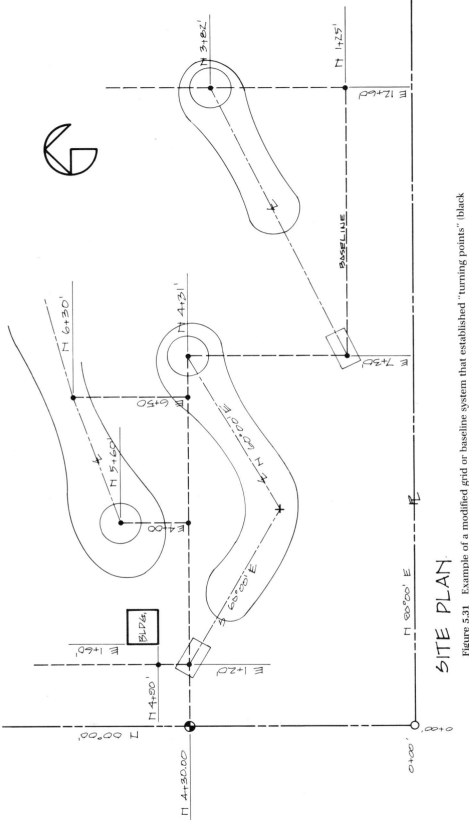

SITE PLAN

Figure 5.31 Example of a modified grid or baseline system that established "turning points" (black dots) at critical points within a site. Note that baselines may be given a bearing that is *not* parallel to the property lines by the use of two coordinate numbers. The system is adaptable to sites requiring only approximate site feature locations.

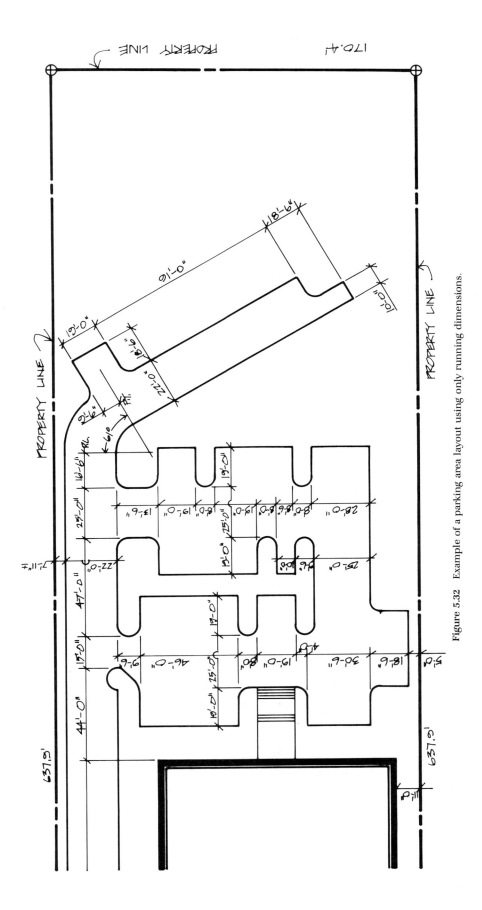

Figure 5.32 Example of a parking area layout using only running dimensions.

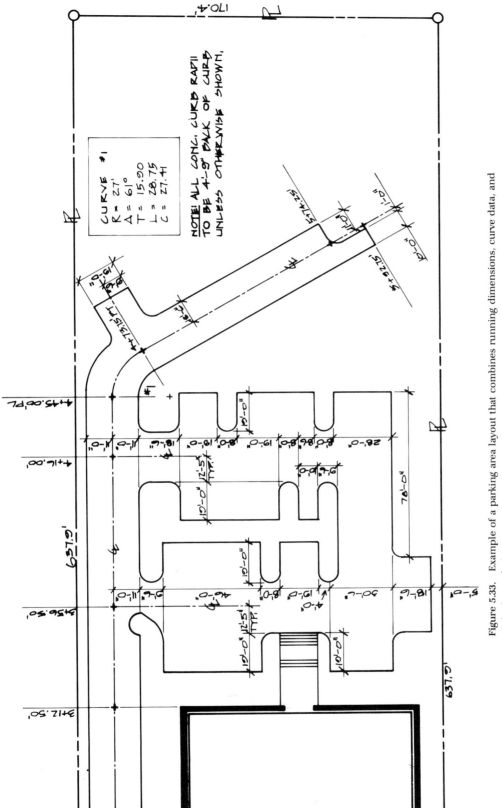

Figure 5.33. Example of a parking area layout that combines running dimensions, curve data, and stationing techniques.

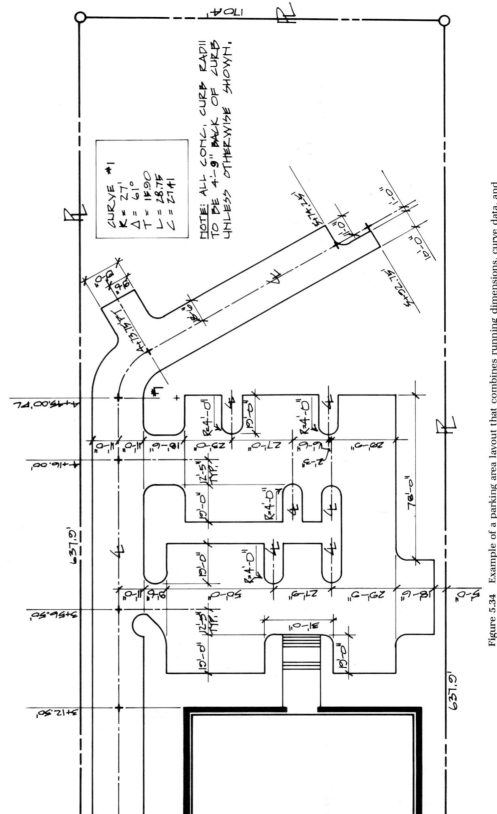

Figure 5.34 Example of a parking area layout that combines running dimensions, curve data, and stationing techniques with predominant use of centerlines.

124

SELECTED READINGS AND REFERENCES

Barry, Autin B. *Construction Measurements*. New York: Wiley, 1973.

Blachut, Teodor J., et al. *Urban Surveying and Mapping*. New York: Springer-Verlag, 1979.

Carpenter, Jot D., Editor *Landscape Architecture Construction Workbook*. American Society of Landscape Architects Foundation.

Parker, Harvey, and John W. MacGuire *Simplified Site Engineering for Architects and Builders*. New York: Wiley, 1954.

Chapter Six
Controlling Earthwork and Conduits

*A society that scorns excellence in plumbing because plumbing is a humble activity
and tolerates shoddiness in philosophy because it is an exalted activity
will have neither good plumbing nor good philosophy.
Neither its pipes nor its theories will hold water.*

John W. Gardner

6.1 THE DRAWINGS AS INFORMATION

A contract document ordinarily includes aspects of grading, drainage, and earthwork in association with site preparation and finish elevations. In many respects, it is helpful to think of these actions and elements as part of vertical control. Vertical control implies design and construction processes of greater proportions than grading of soil. It relates to an interdependence among surface soil, storm water surface flow, subsurface storm water, various subsurface and aerial utilities, walls, buildings, roads and curbs, parking areas, walks, existing site characteristics, and off-site features. Contract documents must relate to the entire design concept by assuming command of those relationships among all vertical components of the design.

Horizontal design aspects are relatively easy to perceive in two-dimensional drawing form, while vertical control calls for exceptional talent in visualizing design and in contract drawing preparation. Vertical control requires symbols that are either quite abstract, such as contour lines, or mathematically precise, such as spot elevations. In most

John W. Gardner, from "Thoughts—on the Business of Life," *Forbes*, Aug. 1, 1977, p. 80.

instances, only the designer's imagination can be useful in previewing the finished earthwork or construction.

The following paragraph is generally found in some form in the conditions to any contractual agreement:

> *Due to the small scale of the drawings, it is not possible to indicate all offsets, fittings, etc., which may be required. Drawings are generally diagrammatic and indicative of the work to be installed. Precise structural, subsurface, and finish conditions are not, and cannot be, furnished to the contractor prior to installation.*

Drawings describing system installations are about as abstract and symbolic as any graphic information prepared by a site planner. As the foregoing paragraph implies, a drawing is diagrammatic rather than precise information. On a diagram, the locations of various systems are shown only in their approximate positions relative to one another and to site plan features. A drafter strives for graphic clarity when preparing such a drawing, that is, each system and its symbolic components are distinct and separate, regardless of their graphically scaled dimensions in relation to one another or other physical features. As a two-dimensional plan, vertical aspects must be portrayed through symbols or spot elevations, or related to large-scale details. Because a diagrammatic drawing is totally symbolic, a precise legend and schedule are critical to the contractor's interpretation and translation of each symbol's meaning.

Written technical specifications will generally interpret and describe each of the various component materials, installation practices, system testing, performance, and guarantees. In many respects, the technical specifications become an extension of the legend and schedule and must be coordinated with them to avoid redundancy and conflict.

Contract document information must often include the horizontal and vertical locations of the pipes. Systems that are to be fixed in their horizontal location in easements must be located with horizontal control data related to the precise survey of the easement. Other types of systems may simply be scaled as to their approximate locations. Whenever the entrance and exit points are to be located precisely, such information must be placed on either the horizontal control drawing or on whichever drawing will control such information. Vertical information is usually given in the form of spot elevations relative to the bench mark. It is common practice to place spot elevations at each change in a pipe's horizontal or vertical deviation from a straight line. Percent of gradient may be given alongside the drawn pipe symbol, but may be redundant information. Any graphic details explaining trench width, backfill material makeup, bedding, or installation must not include trench depth as a precise dimension (because of the variability of actual trench depths).

6.2 DESIGNERS

The fact that several professions may be simultaneously involved in the concept of site planning and preparation of contract documents has been discussed in previous chapters. If each of these professions proceeds independently, ample opportunity exists for an owner to receive less than the sum of the professional work. Although each designer strives for the best, it is not uncommon for system designs to be redundant, overlapping, inefficient, wasteful, and in direct competition for subsurface space.

Systems are interrelated among buildings and the site. For example, an irrigation system will generally relate to the domestic system as its water supply, the electrical system for its automatic controls, drainage systems for the timing of system operations, and perhaps effluent or retention ponds for a water source, and may affect a fire control system's performance during emergencies. In addition, where low-density residential land use prevails, the domestic system may be designed as a major part of domestic pipe sizing and metering devices. During the construction of roads and paving, it may be necessary to plan installation of sleeves or connections for later use by an irrigation contractor. In many instances, the locations of telephone and electrical surface junction boxes and transformers must be known before sprinkler heads and plantings are located.

The solution to designer coordination begins at the time an agreement between the owner and designer is prepared. A designer's responsibilities for coordination of information should be made clear.

6.3 SPECIALITY CONTRACTORS

By definition, speciality contractors are generally single purpose. They proceed in a very diligent manner to complete their phase of the work but are not involved in the day-to-day planning or organization of the work's progress. The nature of their work requires someone to coordinate where they must work and the sequence of their work. Problems develop only when several single-purpose contractors are forced, by circumstances, to work together in an uncoordinated manner. An unwritten rule exists on a job site to the effect that the contractor who gets the pipe in the trench first must be protected from those who come later. If contractors are forced to compete on a first-come, first-served basis, an owner receives what may appear to be a battle-field rather than an orderly site.

The following scenario illustrates what must be prevented. The contractor for whom you are responsible has just stopped work because the machinery cannot function on a steep slope not indicated on the grading plan prepared by someone else. Someone inadvertently ordered the work to proceed in that area before the grading contractor was through with the finish grade. While the grading contractor was working, the equipment struck and broke a water pipe that should not have been there.

Meanwhile, the roof and storm drain contractor cut 30 control wires serving 30 automatic irrigation control valves at the other end of the site. The contractor was searching for the connection to the storm drain that had been installed several days ago. The connection had been accidently covered by the spoil from a contractor who was repairing a broken water main. While the water main was being repaired, the local water department drained its lines to repair an off-site break and siphoned the polluted water throughout the site's water system. The local health department required all installed pipe to be flushed and chlorinated before domestic use. And on and on.

6.4 MULTIPLE CONTRACTORS—COORDINATION

The method of contracting can signal success or chaos for the designer and owner. If an owner or designer allows several prime contractors to enter a site without coordination, conflict and malfunctioning systems will be the rule rather than an exception. At best, coordination is usually marginal, that is, one contractor can accidently excavate, cut, damage, or otherwise unbalance another system at any moment. Left to formulate their own independent procedures and sequences, contractors can find themselves in a ditch trying to splice wires or patch pipes back together. It is imperative that a third party be made available to track installed work and direct the sequence of new work. Whenever a designer is involved in the design of a site system, there must be an understanding from the onset as to the degree of responsibility and coordination that will rest with the designer during installation. Before any agreement is signed between the owner and a contractor, there must be an understanding as to whom the contractor looks for the coordination of work. Both the designer and owner must recognize that a high degree of liability will accompany the direction of work sequences and underground excavations.

6.5 MULTIPLE CONTRACTORS—EXISTING CONDITIONS

Each prime contractor and subscontractor must accept the work of previous contractors. Two options are open to each new contractor. First, existing site conditions can be appraised for what they are rather than as indicated on the drawings and possibly a contingency added because of a poorly defined scope of work. Second, the contractor can refuse to accept the existing conditions because they do not reflect those indicated on the working drawing.

The lesson to be learned in preparing finish grading plans when multiple designers and contractors are involved is as follows:

1. All information should be obtained from contract documents of previous grading operations, and its accuracy determined. Such information will become the basis of "existing conditions" with respect to contract documents being prepared and upon which the new contractor must bid or perform.

2. It should not be assumed that a surveyor's original survey of existing conditions shows those conditions that actually might be faced by a designer and contractor.

3. Contract documents should be prepared in such a manner that a new contractor is protected against having to accept existing conditions that are not described in the documents. The best way to accomplish this is clearly to indicate those conditions that *will exist* at the time site work begins.

4. If things are really in bad shape, the designer should insist upon a new survey of the property as its exists, after previous construction, before proceeding with drawings and defining the scope of work.

5. Contractual options available to a new contractor should be examined. Are the existing conditions undefinable and is a lump sum, time and material, or other type of agreement necessary?

6. The designer should not rely completely on the contractor's normal site visit during bidding as a means of obtaining a complete picture of the site conditions. There is often a considerable time lapse between bidding and an owner's acceptance of a new contractor. During the delay, a site can change so much that the new contractor may seek to void a bid on the basis of changes to existing conditions.

6.6 ACCURACY

Only two forms of information are available to the drafter for the purpose of portraying vertical elevations of land forms. The first are semifixed, abstract symbols that can be used under conditions allowing flexibility in accuracy or that may require on-site aesthetic judgments. The second form includes fixed and mathematically precise numerals called spot elevations. In most instances, both of these information forms are used on the same drawing to identify variations in required accuracy.

6.6.1 Semifixed Controls

Semifixed controls are symbolic. Graphic use of such a control implies that the designer will allow a site's conditions, the availability of soil, or esthetic judgment to dictate the final earth or structure form. Obviously there are times when such flexibility of purpose and contractual obligation is desirable or even necessary, but equally obvious is the potential for indecision and the creation of an indefinable scope of contractual work.

The following illustrate several types of symbolic and semifixed means of representing the nature of contract work.

1. *Symbolic top and toe of slopes.* Figure 6.1 illustrates, symbolically, the approximate location and size of a slope. A designer believes the slope to be necessary but can only approximate its location and size. The symbol can be coupled with spot elevations and/or dimensions for delineation of relative size and location of a slope. If the top and toe locations are not fixed by spot elevations, the symbols can be supplemented by a ratio indicating a slope's minimum or maximum gradient.

2. *Hatchures.* Figure 6.2 illustrates the use of

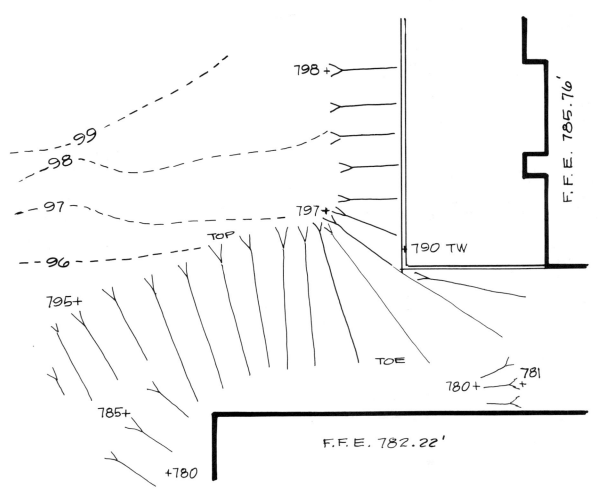

Figure 6.1 Use of symbols to denote the probable configuration of cut or fill slopes.

hatchures with respect to the approximate location of borrow or fill areas. The hatchures may be used alone simply to identify general areas to be graded or can be coupled with specified volumes of cut and fill requirements. Notations may be also useful for indicating volumes, finish slope gradient, and the like. Hatchures are a poor substitute for a contour line because such graphics imply a rough form without precise control of finish elevations.

3. *Contour lines.* When only contour lines delineate a topographic form, the acceptable finish grade can be no more accurate than the contours themselves. Existing contours are generally recognized as accurate to one half of their contour interval. Figure 6.3 can be no more accurate than ± 1 foot of vertical elevation. A designer may argue that a greater accuracy is intended because the proposed contours are an expression of intent, but it is arbitrary to assign proposed contours a greater accuracy than it is possible to attain for existing contours. If greater accuracy is intended, such

intent should be expressed by spot elevations, in addition to or in place of contour lines, or the contour interval of both existing and proposed contours should be decreased by a new survey.

6.6.2 Fixed Control

Fixed control is obtainable with a spot elevation. However, the intended accuracy of a spot elevation may or may not be fixed, depending upon the degree of accuracy delineated on the drawing. A designer's intent is described in much the same fashion as for horizontal control, that is, by the number of significant figures given or the use of a plus or minus designation.

The degree of accuracy required depends on the topography and the construction. Finish floor elevations, tops of walls, paving, and similar features are usually defined in terms of ± $\frac{1}{100}$ foot. A spot elevation is then given to two decimal places. On the other hand, soil surfaces are often subjected to

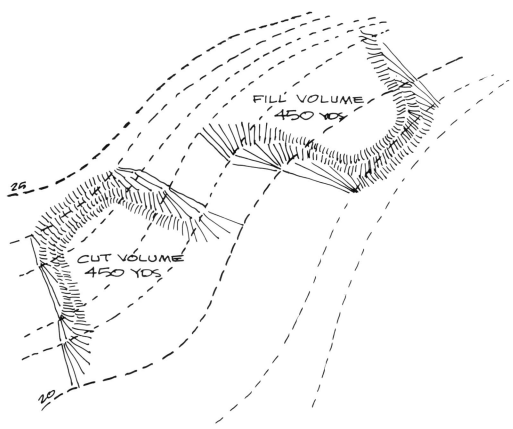

Figure 6.2 Use of hatchures to denote the probable configuration of cut or fill slopes.

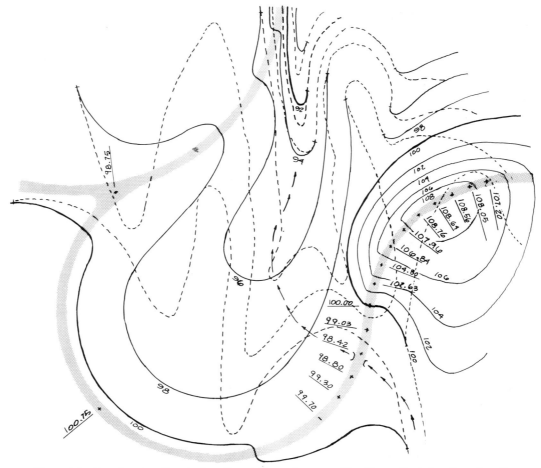

Figure 6.3 Use of contour lines to describe existing and proposed land forms. Spot elevations describe
several vertical curves in plan form; if vertical curve profiles are used, these spot elevations become
unnecessary. Arrows indicate storm overland drainage direction.

vertical expansion and subsidence during and after
grading operations. Finish soil surface elevations are
adequate to about plus or minus one decimal place,
that is, $\frac{1}{10}$ foot.

The term *finish grade* is defined by the contract
documents, and may be defined quite differently by
different contractors. In ordinary usage, finish will
mean the final elevation of land or structures within
the prescribed limits of intended accuracy. However, to a grading contractor who ordinarily may be
involved in rough grading with large equipment, the
term may mean finish subgrade elevations. And to a
landscape contractor, it will mean the final elevation
of the soil surface prior to the planting of seed or
installation of plants. In essence, finish must be
defined in terms of the final contractual obligation of
a specific prime contractor.

How many spot elevations are necessary? The
answer is difficult to deal with because the situations vary with need. Basically the number of spot
elevations shown must match the designer's intended degree of accuracy and the complexity of the
soil's surface configuration. One rule prevails: There
is an implied straight-line gradient between any two
spot elevations. Only two spot elevations are necessary whenever the gradient between them is to be
true to line, regardless of the distance between the
elevations. However, it is common courtesy for the
designer to delineate spot elevations at a spacing
related to that gradient. For example, if the gradient
is to be 1%, delineating a spot elevation about every
25 feet will aid the contractor's grading and the
designer's checking of the work's accuracy. A 1%
gradient is critically marginal in terms of storm

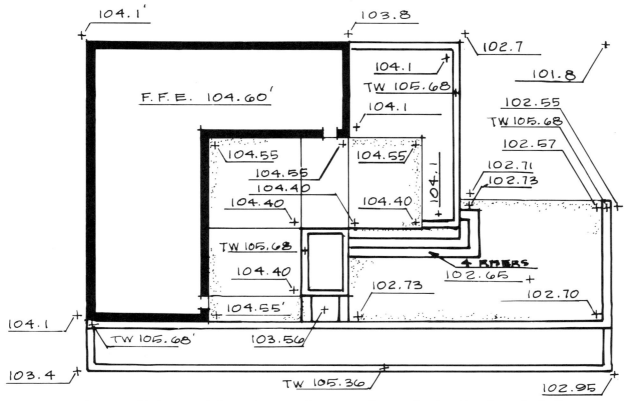

Figure 6.4 Maximum vertical control accuracy demonstrated by spot elevations. Note the two levels of accuracy provided by the number of significant numbers. Top elevations of walls are designated TW. The builder's floor elevation is designated F.F.E. (finish floor elevation).

water drainage and requires accuracy in its construction. On the other hand, a gradient of 10% is not a critical drainage problem and two spot elevations are all that are necessary to check the work.

The number of spot elevations delineated on structural work varies somewhat, but basically relates to the way in which the work is executed and to the materials involved. Poured-in-place concrete, for example, must be formed and spot elevations are required only at abrupt changes in elevations. The top of a concrete wall that is 100 feet long and dead level needs only one spot elevation, but if "stepped," it requires two spot elevations at each step. Concrete slabs are given a spot elevation at each corner if the gradient is uniform and of straight pitch between the corners. Because corners must be located by horizontal control, elevations might as well be found at the same time. If a concrete slab surface pitches or is warped, however, these breaks in gradient must be given spot elevations.

Dead-level surfaces require only a single spot elevation. For example, the floor of a building may be given a spot elevation preceded by F.F.E. (finish floor elevation) to indicate its single, level elevation.

Subsurface pipes are usually designated as to their invert (flow line) elevation. Gravity systems, such as storm drains, require invert elevations at each change in gradient to control the flow velocity and direction. Pressure systems, such as water, require spot elevations whenever the water must be drained each winter. Conflicts may result if pipe invert, top, and diameters are all given in the contract documents.

Whenever possible, the drafter should coordinate horizontal and vertical controls. Spot elevations scattered about the drawing are of little value if they cannot be found by horizontal control. Such a process is not a requirement; it is simply a courtesy to the contractor and increases construction accuracy.

The following indicates where spot elevations may be necessary on a grading plan and their possible coordination with the documents as a whole.

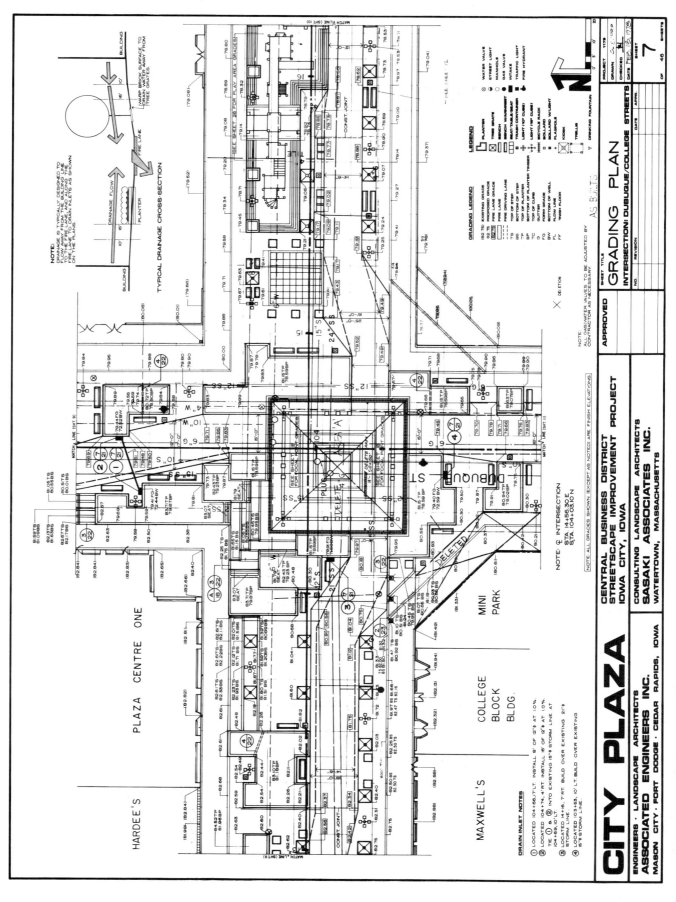

134

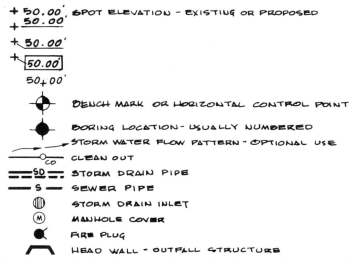

+ 50.00' SPOT ELEVATION - EXISTING OR PROPOSED
+ 50.00'

+ 50.00'

+ 50.00'

50+00'

BENCH MARK OR HORIZONTAL CONTROL POINT

BORING LOCATION - USUALLY NUMBERED

STORM WATER FLOW PATTERN - OPTIONAL USE

CLEAN OUT

STORM DRAIN PIPE

SEWER PIPE

STORM DRAIN INLET

MANHOLE COVER

FIRE PLUG

HEAD WALL - OUTFALL STRUCTURE

Figure 6.6 Example symbols often used on grading or utility plans in order to save drafting time and to allow explanations within limited graphic space. Symbols may vary regionally.

1. Wherever a specific point in vertical and horizontal space requires a more precise definition than is possible with semifixed delineations
2. At the peak and low points of earth forms or paving
3. At the corners of structures that are not to be constructed dead level; coordinate with the design gradient, material drainage and construction characteristics, and horizontal controls
4. As a single finish floor elevation for structures, and elevations of soil, or pavings that are to be constructed dead level; coordinate with storm drainage characteristics if the surface might have drain inlets
5. At the finish surface of each adjacent to the sill plate of structures, or adjacent to walls and fences; coordinate with detail drawings, storm drainage flow, and local construction codes
6. At the top and bottom of any abrupt change in the longitudinal or sectional gradient of a slope, swale, ridge, or the like

7. At frequent intervals along earth surfaces that are to be constructed with less than minimal gradients
8. At frequent intervals along gradients that are to be accurately curved in section or longitudinal profile; coordinate with vertical road curve and profiles as well as with shaped earth berms or swales of esthetic interest
9. At points of utility distribution or collection systems as they relate to the finish earth surface, structures, inlets, invert elevations, outlets, paving, local codes, and functioning of the various systems; coordinate with all system types
10. At the tops and footings of walls; coordinate with construction details and subsurface and surface drainage plans
11. At the tops of piers and footings supporting structures; relate to the finish elevation of any structure being supported, their construction details, local codes, and finish soil gradients

6.7 VOLUME, AGREEMENTS, AND SPECIFICATIONS

The cost of site grading will depend upon the volume of soil, precision of soil placement, and the operations necessary to achieve compaction. Drawings allow both the designer and the contractor to arrive at separate estimates of the soil volume, and

Figure 6.5 An example of a grading plan that depends upon spot elevations. Note that the legend is necessary to define each elevation relative to a particular site feature. Subsurface utilities are indicated so that a contractor may determine the amount of excavation occurring between conduits and finish grades, although other drawings may carry the elevations of conduits. (Courtesy Jack Leaman, landscape architect, Ames, Iowa)

access to the proposed elevations. Technical specifications outline the quality and degree of compaction necessary in the various parts of the project. A contractor's estimate of soil volume and procedures becomes the agreed-upon price for the work, either as a unit basis or a lump sum. A designer usually estimates the grading costs on the basis of the anticipated volume of soil and finish work involving the contractor. The owner should approve a designer's estimate of costs prior to accepting bids for the grading work.

Contract documents and agreements among the parties must recognize the fact that soil, and those procedures necessary for grading and measuring it, are not absolutes. Materials and equipment are inexact and subject to variable conditions beyond the control of a designer or contractor. Attempts to demand perfection in the restructuring of land or the measurement of volumes will lead to frustration. However, much of this can be eliminated by the designer during the preparation of the drawings and the arrangements between the owner and the contractor.

6.7.1 Lump Sum Agreement

Lump sum agreements for earthwork and finish grading operations often seem deceptively simple to a designer. Commonly a grading plan is prepared, bids are let, and a lump sum agreement made between the owner and contractor. Behind such a scene, however, may lie a presumption of information that may later prove defeating to all parties.

The taking of lump sum bids and the subsequent agreement for the work implies that all necessary information is known to the contractor. In many instances, such an assumption is unfounded. Generally, what really exists is a contractor who is gambling on the presumption that sufficient information has been supplied. An experienced contractor will either build contingency funds into the bid or count upon extra compensation from change orders as the unknowns become obvious to the designer and owner. In either event, the contract documents will be constantly searched for discrepancies between the existing conditions on the site and those finish elevations required by the designer. (See Section 3.2.)

Several reasons usually account for major problems as the work unfolds. For example, the base information of existing conditions may not match the precision required by proposed characteristics,

soil characteristics may have changed during the planning and bidding process, information was simply incorrect, or all of these factors may have combined to produce problems.

A mismatch between existing and proposed conditions is a common situation. Consider, for example, an existing condition plan indicating the topography of the site at 2-foot contour intervals. This base information is then coupled with a finish grading plan calling for precise fixed elevations to $\frac{1}{100}$ foot. Chapter 3 indicates that 90% of those contours could vary in accuracy by as much as $\pm$ 1 foot. With the proposed work requiring a precision to $\pm$ 0.05 foot, how was the volume of soil calculated? The answer, of course, is that a contractor must gamble on a presumption of information. With such a mismatch between existing and proposed conditions, the volume of soil, as computed from the drawings, must be in error. As existing topographic condition information decreases in accuracy, and proposed elevations increase in accuracy, the mismatch increases and the potential accuracy of a contractor's price or the designer's volume estimate decreases.

The problem of mismatch is often further compounded when technical specifications simply imply that a contractor is to bring in fill or haul off site any soil that does not allow attainment of the proposed elevations shown on the drawing. When a mismatch exists, a contractor cannot accurately predict volume and therefore cannot predict price. A contractor's only remedy is to estimate a contingency factor relative to the unknown quantities of cut or fill that may or may not be required to carry out the proposed work.

A designer may inadvertently compound problems by releasing or otherwise implying that certain volumes of soil are a part of the contract agreement. When a contractor is functioning under a lump sum agreement, a designer must not report or imply quantities in the contract documents. Documents that contain *both* volume and proposed site elevations produce two scopes to the agreement. For instance, assuming that a designer's estimate of volume falls short of the necessary volume required to achieve proposed elevations shown on the drawings, is the owner obligated to pay extra compensation for imported soil or are the proposed site elevations to be adjusted to fit the volume of soil on the site?

The following are a few steps that may be taken during the preparation of lump sum contract for grading of a site.

1. When precision finish grades are necessary, match the existing topographic conditions as closely as possible.

2. Maintain clarity of responsibilities by not revealing any portion of a designer's estimate of soil volume unless the lump sum base bid is for a specific and measurable volume of soil; any authorized importation or exportation of soil relative to the site is paid for by firm unit bid prices and noted in the agreement; or provision is made in the documents for the adjustment of proposed elevations in relationship to the actual volume of soil found on site.

3. Reduce legal and price contingency factors by revealing all known information about the topography and subsurface conditions of the site; however, control warranty of such information.

4. Eliminate from the technical specifications all "if" and "maybe" statements, for example, "if fill is necessary", "if cut is necessary," "compaction may be necessary," "imported soil may be necessary," or "if soil must be removed." An "if" or "maybe" statement generally indicates a contingency factor somewhere in the documents.

5. Whenever a contractor is allowed to adjust finish grades in relationship to the volume of available soil, check for potential storm drainage problems, potential effect on building finish floors and local codes, a distinct graphic separation of those spot elevations that may or may not be adjusted, the relationship among finish grades and paving or walls, designer approval or setting of adjustments to finish grades, potential effect on depth of soil cover over utility or irrigation system components, structural effect on supports for exterior decks with fixed finish floor elevations, and planting soil depth.

6.7.2 Unit Price Agreement

Whenever unit prices are used, less emphasis need be placed on the relationship between contractural information and the method of payment to a contractor. However, more attention must be given to the administration and measurement of each unit. A contractor may give a unit price for time with specific grading equipment to execute the work. When the units are in yards, there exists an obligation on the designer's part accurately to measure the volume of soil actually worked by the contractor. It is also important to identify the several types of "yards" that may be encountered on the site and which may be paid for as separately priced units.

A unit of soil is commonly measured as a cubic yard ($3 \times 3 \times 3$ feet, or 27 cubic feet). Unfortunately there are several types of yards, which vary in their definitions and in their unit prices. Drawings, technical specifications, and particularly the bid form must define the type of yard being discussed and how it is measured. The following are examples of common construction definitions. It is always best that contract documents define precisely what is intended by the use of each term.

1. A *bank yard* is 27 cubic feet of undistrubed soil in situ. When cut volume is computed from a drawing, it will be a bank yard.
2. A *cut yard* is one bank yard.
3. A *loose yard* is one yard of material measured immediately after it has been loosened from an in situ position. See truck yard.
4. A *truck yard* is one yard of soil or material as it exists in the bed of a truck or in a loosened condition. One yard will contain voids that will partially disappear with vibration or compaction.
5. A *compacted yard* is one yard of soil measured after being mechanically compacted to the density prescribed by the technical specifications.
6. A *fill yard* is one yard of soil moved into an area, and may be measured in a compacted or loose state. It is subject to various regional definitions. When measured from a drawing, a fill yard is assumed to be in a compacted state.

6.7.3 Finish Terminology

Regional differences in term definitions exist. The following definitions will serve for purposes of technical specifications.

1. *Finish or fine grade* is a completed soil surface brought to the topographic accuracy expressed by the drawings. It is commonly considered the last major phase of grading operations but does not necessarily specify texture. For instance, is the finish texture devoid of abrupt changes in line, stones, clods, rivulets, tire marks, and the like, or may the surface be of a rough texture, unfinished, with stone and clods, grading equipment tracks and marks, or survey pits or mounds? In many regions, this grade will be that which is achieved by adherence to the "blue top" surveyor-set grade stakes. The

term *fine grade* is also commonly used to describe the *action* of smoothing and raking the soil surface prior to planting or seeding plant materials.

2. *Rough grade* is a soil surface that has been brought to near finish grade. Normally slight cutting and filling by small equipment is required to establish the finish grade. Whenever multiple contracts exist, rough grade must be defined in terms of acceptable accuracy with respect to plus or minus fixed elevations. Completion of the rough grade by one contractor presumes that no importation or exportation, or lengthy transportation, of soil will be necessary for the establishment of finish grade by another contractor.

3. *Subgrade* is the final elevation of soil upon which paving or paving subbase material will be directly placed. This is a common source of misunderstanding among multiple contractors in that contract documents may or may not clearly define which contractor is to be responsible for establishing the subgrade.

6.7.4 Soil Terminology

The following terms, often found in technical specifications relating to planting and grading work, influence the cost estimate and units in a bid form.

1. *Shrinkage* is an estimated percentage of variation between a bank yard and a compacted yard. It will vary from 10% to 30% of the estimated volume of compacted fill. Shrinkage is usually involved in estimating the value of a contractor's work. For instance, the shrinkage factor accounts for an estimator allowing 10% to 30% more cut than fill because (a) soil may be compacted to a density greater than occurred *in situ*, (b) soil is "lost" during transportation and movements, (c) errors arise in finished earth forms, and (d) compaction may be greater than the minimum specified.

2. *Fluff* is the opposite of shrinkage. The term is applied generally to estimating the number of truck loads of soil necessary to achieve a specific compacted yard. For instance, if the loosened yard exceeds an in situ yard by 15%, it is said to *fluff* 15%.

3. *Select fill* is material that has been approved for use as fill and which meets or exceeds the physical and structural specifications for suitable fill.

4. *Undisturbed earth* refers to soil in situ, that is, not loosened. The term generally is used to warn a contractor not to disturb the earth beneath footings,

piers, and other load-bearing elements that require a bearing on undisturbed soil.

5. *Top soil* is as defined in Chapter 7.

6. *Borrow pit* is a place from which soil may be cut and transported to a site or elsewhere within the site for use as fill material. If a borrow pit is within the site or otherwise known and acceptable to the designer, the pit's location is delineated or described in the contract documents. When and if a borrow pit is identified, the documents should specify the finish topographic form of the pit and any managment controls that may be required of the contractor when finished with the pit. Specifications must also identify responsibility for dewatering of the pit, if necessary, during or after the grading operations are complete. A contractor *must* presume that borrow taken from a designated borrow pit is suitable fill material.

7. *Borrow* is material removed from a borrow pit.

8. *Spoil* is excess material cut from the site and either transported from the site or placed in areas as fill and as described in the contract documents. The contract documents must describe the finish grade and configuration of the deposited spoil material. Spoil may be placed as permanent fill or temporary stockpile.

9. *Stockpile* is an area described on the drawings as suitable for the temporary storage of excess soil material until it is incorporated into grading work.

6.7.5 Significant Change in Volume or Scope

A significant change in the scope of construction work may be grounds for nullification or adjustment of an agreement. A significant change is a condition that revises the terms of an original agreement in favor of one of the parties. It is not uncommon for arbitration or the courts to hold that a contract agreement is void whenever conditions affecting the scope of work have significantly changed.

Both lump sum agreements and unit price agreements are affected by a significant change in the scope of the work. Lump sum agreements may be affected to a lesser degree because of a contractor's presumed acceptance of greater risk, but unit price agreements often place a great deal of responsibility on the designer. For instance, a contractor's unit price is based on the quantity of units. If the quantity of units significantly changes due to an error in the designer's quantity estimate, a contractor may recover damages from an owner, revise the unit price, or void the agreement.

When a grading contract is let under a lump sum agreement, the contractor is solely responsible for estimating the scope of work. The contractor is usually dependent upon drawings as a means to determine the volume of soil necessary to establish proposed finish elevations. Ordinarily, such a common practice is without problems unless a significant error in the drawings is responsible for a significant change in the volume of soil required to complete the work.

If a contractor can prove that drawings were in error—that is, they misrepresented the scope of work—extra compensation may be requested. At issue will be the amount of risk accepted by the contractor at the time the agreement was signed. For example, if no drawings existed, a contractor would accept total risk of estimating volume and scope. On the other hand, when drawings exist, the owner, through a designer, is accepting a portion of the risk by making documents a part of the scope of the work.

Unit price arrangements are commonly used whenever existing or proposed information does not allow a firm estimate of soil volume. In effect, an owner is accepting a greater share of risk than would be the case with lump sum agreements. Ordinarily a designer's estimate of volume will become a part of the contract documents and agreement. The owner, through the designer, is thus establishing the scope of the work through drawings of proposed and existing conditions and the volume of soil required to complete the scope of work.

The degree of risk accepted by an owner and contractor will depend upon the type of unit price and means of measuring each unit. If, for instance, a contractor is to be paid in units of time, the contractor accepts very little risk, except for the price per hour of time. On the other hand, if the cost per unit is to be based on a measured-in-place compacted yard of fill, the contractor must accept the risk of efficiency in time as well as price per yard. From a designer's standpoint, drawings must be more accurate when units are for earth in place than for time expended by a contractor.

A designer's role is to advise an owner of the relative risks involved in each type of unit pricing. It is the owner's decision to accept or reject risk and a designer's job to express such a decision through the contract document format. Each contractor must decide, in turn, whether or not the risks expressed in the contract documents should be accepted.

6.8 PERFORMANCE AND PROCEDURAL SPECIFICATIONS

The language of a grading specification and graphic technique may be either performance or procedural in nature. A performance format fixes and objectively measures the end product of the contractor's scope of work. A language or a drawing that indicates proposed fixed and semifixed elevations, drainage patterns, the structural character of in-place soil, and the finish surface of the grade constitutes a performance-type document. A language or a drawing that directs procedures or equipment to be used by a contractor is procedural in nature. A performance type of specification, for instance, states certain tests to be performed on the contractor's finished work. The contractor warrants compliance with the specifications and is paid only upon successful completion of tests (lump sum or unit price agreements). Procedural-type specifications will dictate, for example, the type of equipment to be used, the number of passes over the area, and the method of obtaining a fine grade prior to the seeding of grass.

The distinction between procedural and performance languages is purely academic. At issue is the effect each type of specification may have upon relationships between the contractors and the owner. From a designer's standpoint, a contractor and owner accept responsibility for the finish products of the work in relation to the specification language. Exactly what those responsibilities will be and which person will accept them must be kept firmly in mind during the development of specification language. A procedural language, for instance, appreciably lowers the risk taking of the contractor and shifts greater risk to the owner. A performance specification places the total burden of risk on a contractor, at least to the extent prescribed by the objective tests that evaluate performance. Under procedural language, a contractor does not warrant a successful product, but only faithful compliance with the procedures directed by the specification language.

To avoid an owner's obligation for performance, some specifications attempt to develop both procedural and performance languages. For example, the following two paragraphs have been taken out of context to illustrate how a designer has specified certain procedures but has directed the contractor to submit alternative procedures. If the contractor does not submit alternative *procedures*, the language presumes that the contractor will accept full responsibility for the *performance* of the work. It is further presumed that if the procedures fail to achieve the desired compaction, the contractor will perform more work until the compaction is acceptable, at no extra cost to the owner. A contractor is thus placed between the proverbial rock and a hard place.

> *Type 1 compaction refers to compaction requiring a minimum of one rolling per inch depth of each lift, and it is further required that the roller continue operation until it is supported on its tamping feet, or the equivalent.*
>
> *If other types of compacting equipment are used ... compaction will be considered suitable, if the resulting density, with adequate moisture, is reasonably uniform throughout the compacted life and is at least 95 percent of the maximum density, determined in accordance with ASTM D-698 (Standard Proctor Density).*

Under such conditions, it is not clear exactly what is expected of the contractor. What is clear is that the designer did not direct the method without reservations, left alternative methods and equipment to the contractor, and probably confused responsibilities among several contractors and the owner. In essence, the designer is saying, "Follow my directions to the letter, but if they do not work, try something else until performance is equal to what I intended it to be." It may have been best to have specified only the performance test and to allow the contractor to develop methods to achieve the performance required.

This is not to say that procedural-type specifications are not necessary. In common usage, procedures and methods are often required to illustrate the intended finish and fine grading needed to prepare a seed bed, to incorporate fertilizers or soil amendments, to scarify, to prepare a site, and other site grading actions that do not have quantitative or objective tests for compliance and procedures suggested by initial soil tests. However, it is only fair that both the owner and designer stand behind a procedural specification and accept the results of those procedures. When writing a procedural specification, the designer should be aware of potential implications for the guarantee of plant materials, seed germination, drainage characteristics of the site, and relationships to other prime contracts.

6.9 SOIL COMPACTION

Most sites will involve cutting and filling of soil to achieve the proposed grades. Except for unstable types of soil beneath structures, most grading operations will require compaction of only the fill portion of the work. The technical specifications generally carry either procedural or performance information relative to compaction of fill. The previous section discussed the relationships between procedures and performance language. This section concentrates on the terminology and means of testing for compliance with test criteria.

6.9.1 Procedural Compaction

The following terms are often employed in describing procedures. In general, they describe the type of equipment to be used by a contractor and the number of passes over the soil required of such equipment.

1. *Wheel pack.* Compaction is achieved by the rubber wheels of trucks or scappers used to haul fill into a specific area. Compaction will vary with equipment weight, tire size, and soil moisture content. It is difficult, if not impossible, to predict uniformity of compaction. This procedure is useful for filling depressions, for nonstructural loads, as a cheap method, or for temporary storage of materials. It is useless for constructing surfaces that must remain permanently without subsidence, for trench backfill compaction, for slope or embankment construction, and for areas with subsurface utility or irrigation systems.

2. *Water compaction.* Loose fill is compacted by jetting, flooding, or sprinkling to settle soil. The method is suitable for nonstructural fills that will not be subjected to surcharges or other loads. Jetting tends to segregate fine grains in the soil and to damage percolation as well as the general horticultural character of the soil. This procedure is useful in sands but useless in clays and in compacting trench backfills. Sprinkling is suitable for settling tilled areas prior to seeding if loosened soil is less than 8 inches deep. It sometimes is used under controlled conditions for compacting sands with vibration.

3. *Rolling.* The weight and number of passes of a roller may be specified for nonstructural com-

paction. Rolling is useful in compacting fine or tilled areas prior to seeding and in compacting the surface of 6 to 12 inches of structural fills and the finish surface of subgrades. It may be necessary to compact the interface of grubbed in situ soil prior to the placement of fill.

6.9.2 Performance Compaction

The following terms and procedures are in general use for specifying performance characteristics of filled soils.

1. *Relative density.* The specification language term "percent of relative density" is the product of:

$$\frac{\text{maximum weight of soil}}{\text{compacted weight of soil}} \times$$

$$\frac{\text{compacted weight of soil less loose weight}}{\text{maximum weight of soil less loose weight}}$$

A soil's maximum weight is determined by uniform laboratory procedures. The compacted weight is taken as the density of the compacted soil in situ. A soil's loose weight is its density in an uncompacted state, that is, the maximum pore space possible in the soil. Each weight of course, would be based on an equal volume of measure.

In essence, such a specification is an objective measurement of how well the soil was compacted relative to its potential for compaction (100%) and physical characteristics as a soil. In the main, the term and method are useful in controlling compaction of sands. Commonly the relative density values will run from 70% to 85%.

2. *Percent of maximum density.* This is commonly used as a guide to structural quality of mass grading and fills. It is generally specified as, for instance, "compacted soil to meet or exceed 95% of the maximum density as determined in accordance with ASTM D-698." Basically the test is controlled as a percent relationship between the density of a soil compacted in situ and the potential density it is possible to achieve under controlled and uniform laboratory tests. The designer should, be aware however, that such a specification indicates only

compaction and does not necessarily indicate a soil's suitability for bearing structural loads.

6.9.3 Compaction Testing

There are two basic methods for arranging laboratory facilities for compaction testing. Perhaps the more common is for the owner to hire an independent laboratory to field check compaction progress, take specimens for testing, report test results, and recommend procedures to correct any problems. If the owner retains a laboratory, the specifications should so state, as well as, if possible, the types of tests that will be run by the laboratory. A specification commonly will define the method of each test (ASTM, ASSHO, or other), the number of tests, minimum standards to be performed by the contractor, backcharging of the contractor for the costs of tests that fail (optional), and the contractor's provision of access to the site by consultants. Specifications must also describe measures necessary to correct any deficiencies brought to light by the failure of tests.

On occasion, an owner may allow a contractor to hire an independent laboratory to provide the same services as noted above. The specifications must then describe the types of information and tests required of the laboratory and note necessary presentation of information to the owner or the owner's representative.

6.9.4 Compaction and Conflicts

The percentage of compaction often conflicts with site use and requires cooordination among designers and contractors. For example, if one designer specifies 95% compaction over an entire site, those areas of plantings ordinarily must be scarified and tilled to provide a horticultural environment. If the designer who controls the major grading specifications can separate fill areas into variable compaction zones, each zone may be compacted to only that density necessary for its ultimate use. For example, greater than 95% of maximum density is ordinarily suitable for the support of buildings and other structures; 90% is suitable for the support of ordinary slabs on grade, drives, and the like; and horticultural areas will require 85% or less compaction. Particular care should be taken to avoid severe compaction of slope faces to be planted.

6.10   ENVIRONMENTAL IMPACTS

Although long advocated by many designers, attention to the prevention of environmental impacts on and near construction sites is a relatively recent part of contract documents. Beginning slowly in the early 1950s, and accelerating since the 1960s, local, state, and federal environmental concerns have brought the construction site under close scrutiny. Seldom will any major urban site work be executed today without some regulation qualifying site preparation, grading of earth, movement of water, air quality, water quality, control of accelerated erosion, sediment control, and general pollution controls. In the main, all parties to a site work construction contract will become involved, either in securing permits, in observing regulations, or with design considerations, construction methods, or maintenance procedures.

From the standpoint of contract documents, an owner's agent must ensure that clear obligations fall upon the contractor for ongoing attention to regulations and that an owner is made aware of regulations and ongoing maintenance procedures.

6.10.1   The Permit System

A designer may be involved in preparing an overall plan of environmental controls that is acceptable to a local agency. In many instances, securing of approval for a grading permit may be the responsibility of the owner rather than of a contractor. Permits may be necessary from departments of environmental resources, departments of water resources, planning agencies, and local zoning agencies. Such a plan might might also be incorporated in an environmental impact assessment.

6.10.2   Designer Obligations

The initial agreement between owner and designer should carefully identify the role of a designer in preparing the initial plan for the securing of any permits, incorporating and cost estimating work in contract documents, and observing ongoing site work, and that of the owner in evaluating maintenance activities after the construction phases have been completed.

A typical plan and contract document might incorporate a detailed analysis and implementation framework and include the following:

1. Topographic site features
2. Types, depths, slope, and areal extent of soil in a site
3. Proposed modifications to the site
4. Projected volume and peak storm runoff before and after site modification, in conformance with acceptable storm duration and statistical periods
5. Hydrologic and watershed data encompassing the site
6. Sequential staging of earthwork operations
7. Permanent and temporary measures for long-term and short- (construction) term control of erosion, sediment, dust, noise, chemicals, and similar conditions, both on and off site
8. Long-term maintenance program including disposal of entrapped pollutants, control measures, water channels, inlets and outlets, plant materials, and storm water removal

6.10.3   Contract Document Formats

Contract documents can have two basic formats. First, they can incorporate graphic and technical descriptions of exactly what a contractor is to construct, install, provide, remove, schedule, sequence, and maintain during the course of the contract. Second, they can describe, in performance and objective language, what a contractor may be required to prevent and to comply with during the contract's course. The first method presumes two things: that an owner wishes to accept risks for weather, procedures, methods, and compliance with any permit requirements, and that the owner's representative has correctly interpreted and contractually incorporated all criteria associated with any permit requirements. In a sense, a contractor can set a price, either fixed or unit, based pretty much upon precise knowledge of the scope of work. Use of the second method implies that the owner wishes to comply with any permit requirements but hopes to shift most risks of compliance to a contractor. A contractor must then gamble on the weather and the ability to develop methods and procedures as necessary too comply with permit criteria. However, in many instances, an owner may not be able completely to shift compliance responsibility to a contractor owing to the language of the permit or ordinance and, in some instances, of bonding obligations.

It would seem prudent for a designer to develop both procedural- and performance-type documents that vary in degrees of emphasis in accordance with local permit requirements or regulations. Whenever a local authority expressly details procedures, a designer may contractually require those procedures without the ordinary restrictions placed on control of a contractor's sequence of work. Regulatory conditions may expressly require, for example, the procedural backfill and closing of trenches, planting, specific control measures, and minimal exposure of areal soil to be accomplished within certain sequential periods of time. Under performance criteria, a contractor might be held accountable for locating, sizing, constructing, and maintaining certain temporary controls in accordance with the intended and expressed objective of the permit or regulations. At issue is whether or not a contractor had reasonable contractual latitude in developing the temporary means and methods to comply with the objectives of a regulatory agency. Such language should allow a contractor to control methods and procedures while work is in progress, use personal skill in meeting regulatory provisions, and, if required, construct permanent environmental impact controls in compliance with the contract documents. However, it may also be reasonable, under certain conditions of regulation, to develop a "hold harmless" or "liquidated damages" clause that will indemnify an owner for a contractor's failure to meet the conditions of performance expressed by the owner's permit. In other words, a contractor's failure to perform to the regulated conditions may cause a regulatory agency to assess penalties against the owner.

Under some conditions, it may be in everyone's favor to place temporary controls under unit prices within the bid form. As the work of grading or excavation proceeds, the owner, contractor, and owner's agent can assess specific situations of concern and order construction or removal of specific control measures on a unit cost basis. No doubt, such a procedure will place the burden of risk on the owner. However, in a majority of insances, failure to provide sufficient control of accelerated erosion, dust, water quality, debris, sediments, and so on, initially will burden an owner with being in violation of permit regulations, regardless of a contractor's action or inaction.

During the process of permit acquisition and planning, a considerable quantity of information may be gathered and evaluated for on- and off-site conditions. Such data should be handled contractually as discussed in Chapter 3.

6.11 SURFACE CONDITIONS AND ROUTING

Three conditions may conflict, yet be closely related, during the preparation of contract documents. The three—esthetics, accessibility, and costs—are, at times, in direct conflict. However, any designer who cannot find a balance among these issues will achieve less than a successful product.

Esthetic conditions concern the visible location of subsurface system accessories and appendages that obstruct movements, impact upon the safe use of an exterior space, or simply intrude upon the esthetics of that space. An obstruction, such as a telephone splice column, electrical panel, or transformer, may force an area's redesign for pedestrian movement or it may interfere with other systems, such as a sprinkler head. Box or manhole covers can be esthetially out of character with surrounding pavings, as well as a safety hazard if the paving fails or subsides over time (subsidence seems always to occur at some point). Open grates are of particular concern where they can catch bicycle wheels, wheelchairs, or crutch tips.

Accessibility to subsurface system appendages is most often a long-term maintenance and operation condition. In addition, many accessibility conditions involve legal and/or human use of certain zones of a site. Legal conditions arise whenever a system is located in a public utility easement. Long-term accessibility to the surface appendages is a necessity and right for most easements and may involve considerable negotiation with a public utility as to easement location and accessibility rights. In many respects, a site plan will become a function of the public utility rights, points of access, surface planting locations, and pattern of surface land use.

Accessibility to subsurface appurtenances may preclude the surface use of land. For example, while it may be economical to locate sewer cleanouts of water shutoff valves below an area to be used for commercial food sales, it could also pose a financial and health problem to an owner each time a crew must have access to these operational control points.

Planned unit developments may also suffer damages to plantings and turf areas during a public

utility's access to subsurface systems. For example, many such developments are designed for utility routing through open spaces as well as for the more traditional street routing. In most instances, a utility company is allowed access to maintain the system, even though the homeowners' association must repair ordinary damages at its own expense. A designer must not assume that a public utility will be concerned that its 10-ton trucks are creating deep ruts in the turf during an emergency or even ordinary system maintenance. Although many of these issues are settled during the programming and design development phase, many specific decisions remain to be made during the preparation of contract documents.

Daily operational functions relate directly to accessibility conditions. For example, irrigation systems that are manually controlled must be accessible to the operator without the operator being drenched by water from the system's operation. Trees and other major obstacles must not be located so as to preclude the infrequent access of trucks and cranes to lift heavy equipment, pumps, and the like. Subsurface systems that require seasonal or frequent access by trucks or other equipment should be located in clearings so as not to disrupt human uses of a particular place.

6.12 SYSTEM LOCATION

Much of the previous discussion relates directly to many reasons for route selection. However, other design considerations remain. One problem with contract documents is their tendency to become fixed in only two dimensions. As a contract document, the drawings must rely heavily on a drafter's ability to think three dimensionally and a contractor's ability to interpret instructions and information three dimensionally. Unfortunately neither of these presumptions is always correct. As a consequence, what might appear to be simple in diagrammatic drawing form may turn out to be a source of confusion when it must actually be constructed or installed beneath the soil's surface.

6.12.1 Gravity Systems

Systems that depend upon gravity to carry, for example, storm waters, effluents, and subdrainage deserve the highest priority for subsurface space and relationships to finish gradients. Essentially gravity systems are open to the atmosphere and are fixed in their vertical dimensions, that is, there is very little latitude in their departure from dead level and required flow velocities. In addition, their relationship to a site's finish grade must be one of minimal excavation with very little allowable deviation from a straight-line slope. Also, most of the traditional pipe materials function only with precise horizontal angles of pipe entry into a limited selection of couplings and fittings. Some flexibility exists in the horizontal routing of flexible pipes, but they also remain fixed to a straight-line vertical gradient. In all instances, gravity systems are tied to a relatively fixed location of entrance and discharge points.

6.12.2 Pressure Systems

Systems that carry water and gasses are ordinarily pressurized by either a hydrostatic head, a pump, or a pressurized tank. In all instances, these systems function as closed conduits with internalized and variable pressure as well as velocities. A lower locational priority may be placed upon these systems owing to their ability to overcome deviations in both their horizontal and vertical alignments. However, a major design effort is involved in order to maintain certain required horizontal and vertical separations with respect to health and safety conditions, as well as holding minimal depths below the frost line in certain regions. In many respects, these systems tend to follow the finish gradient of the soil's surface, except in cold regions where it may be important to ensure seasonal drainage of the pipes if they are installed above the frost line (pipe is then handled as gravity system).

6.12.3 Electrical Systems

Electrical systems will run the gamut from major power necessary to operate facilities in a building to the installation and arrangement of low-voltage

Figure 6.7 Examples of storm drain underground system. Reference numbers are used in the legend to direct the contractor to a specific detail (on another sheet) that matches the symbols used on the plan view. Note that each drain inlet is given an invert elevation by schedules A and B, pipe sizes are indicated along each pipe symbol, invert elevations are given at intersections and clean-outs, and the pipe material is typically (typ) vitrified clay pipe (VCP). When each drain inlet's grate elevation changes, each inlet grate also needs a spot elevation *in addition to* the invert elevation.

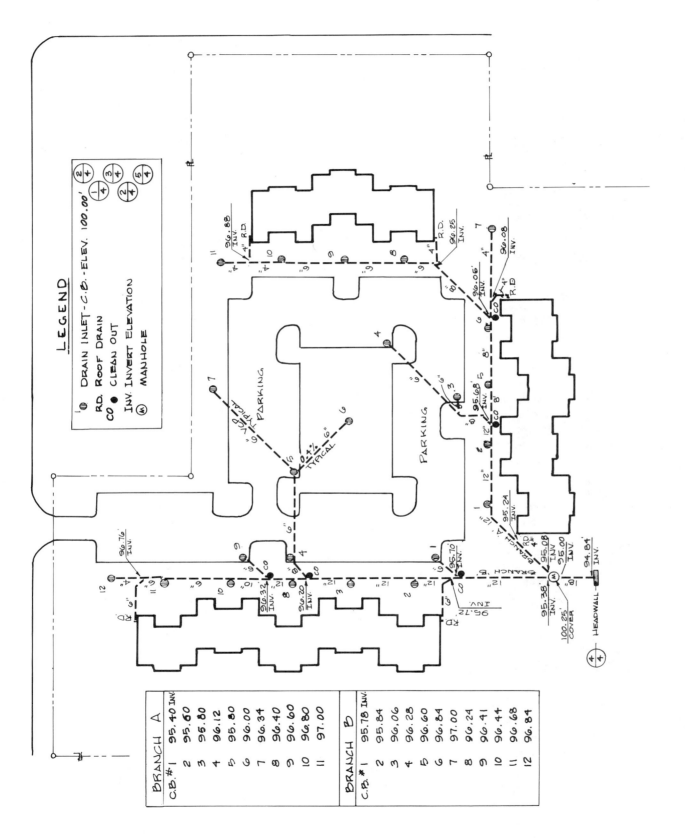

LEGEND

ⓘ	DRAIN INLET - C.B. - ELEV. 100.00'
R.D.	ROOF DRAIN
CO ●	CLEAN OUT
INV.	INV. INVERT ELEVATION
Ⓜ	MANHOLE

BRANCH A	
C.B. #1	95.40 INV.
2	95.60
3	95.80
4	96.12
5	95.80
6	96.00
7	96.34
8	96.40
9	96.60
10	96.80
11	97.00

BRANCH B	
C.B. #1	95.78 INV.
2	95.84
3	96.06
4	96.28
5	96.60
6	96.84
7	97.00
8	96.24
9	96.44
10	96.44
11	96.68
12	96.84

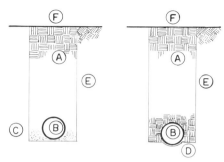

Figure 6.8 Cross section of a typical conduit trench. A designer must decide, delineate, and specify answers to each of the following options when preparing contract documents: (1) character of backfill material; (2) conduit type, size, gradient, fittings, installation, strength, guarantee, testing; (3) character of bedding, granular, concrete, concrete encasement; (4) shaping of trench bottom relative to installation quality; (5) depth of conduit below finish soil surface grade—spot elevations on drawings and/or minimal depth to top of pipe below finish grade (check potential conflicts among grading plan, specifications, and sectional details); (6) maximum width of trench relative to pipe strength and contractor work space.

garden lighting. Each system is commonly quite flexible as to depth, routing, and vertical alignment. However, electrical systems must relate directly to local codes that govern minimum depth, materials, connections, and above-grade appearance.

From the standpoint of site planning, electrical system drawings should be diagrammed in a fashion similar to that for pipe; that is, the drawings delineate the location of conduit or direct burial wire along a fairly precise route rather than in an abstract manner. Architectural electrical diagrams will generally use an abstract routing method; wiring will be delineated as "jumping" from switch to fixture to fixture. When such an architectural abstraction is used in site planning, the designer and contractor will lose the conduit's or wire-routing relationship to other subsurface systems.

6.13   PERFORMANCE AND PROCEDURAL DOCUMENTS

In many respects, utility systems can suffer through imperfect contractual obligations in a fashion similar to those affecting the grading of land and plantings. All three subjects tend to straddle, weave, and vacillate between procedural directives and performance demands. The cause of such confusion seems to lie in the inability of contract documents to deal effectively with unknown site factors. If one couples uncoordinated designers, unattached con-

tractors, variable weather conditions, and the mysteries of a site's subsurface character, it is no wonder that a little chaos slips into the best of relationships.

The relationship of a contract agreement's guarantee and the use of procedural language holds as true for utilities as for plant materials, as discussed in Chapter 7, or for existing conditions, as in Chapter 3. In essence, a contractor who must function under procedural language is not, ordinarily, also held responsible for the short- or long-term functioning of a system.

Procedural nonguarantee contractual formats depend upon an adequate test of workmanship. For example, a section of sewage or storm drain might be tested by methods that measure no more than 200 gallons of leakage per mile of pipe. In this manner, an owner can control the quality of pipe material but may rely upon the contractor and pipe manufacturer for methods of installation. However, if the specifications go too far—that is, they control methods of installation—a contractor may be able to prove that owner directions were at fault for failure of a test. In effect, procedural nonguarantee shifts the burden of workmanship and installation performance to a contractor, within the limits of performance tests, while the owner accepts the responsibility for material quality and system function.

Contract documents may be considered unfair if, whether implicitly or explicitly, they transfer the burden of system performance from a designer/owner to a contractor. For example, some designers may wish to shift the responsibility for sprinkler system effectiveness to a contractor through instructions and guarantee language. Directives might state, in effect, that a contractor is to remedy erroneous documents by adding, relocating, and repositioning sprinkler system heads as required by field conditions. However, even if the owner pays for approved revisions to such a system, an attempt to hold a contractor liable for following documents, rectifying poor functioning, and then guaranteeing system operation and coverage may prove to be indefensible as an aggregate set of contractual obligations.

Figure 6.9 Typical plumbing and surface irrigation plan. Note the use of a legend to identify meanings. See Figure 6.10 for example details. Also note the critical need to use exactly the same language in the legend and to title each detail because no reference symbols exist in this example. It might be clearer if a reference or legend tie to details was used in a fashion similar to Figure 6.7, but the choice depends upon the project's complexity. (Courtesy Jack Leaman, landscape architect, Ames, Iowa)

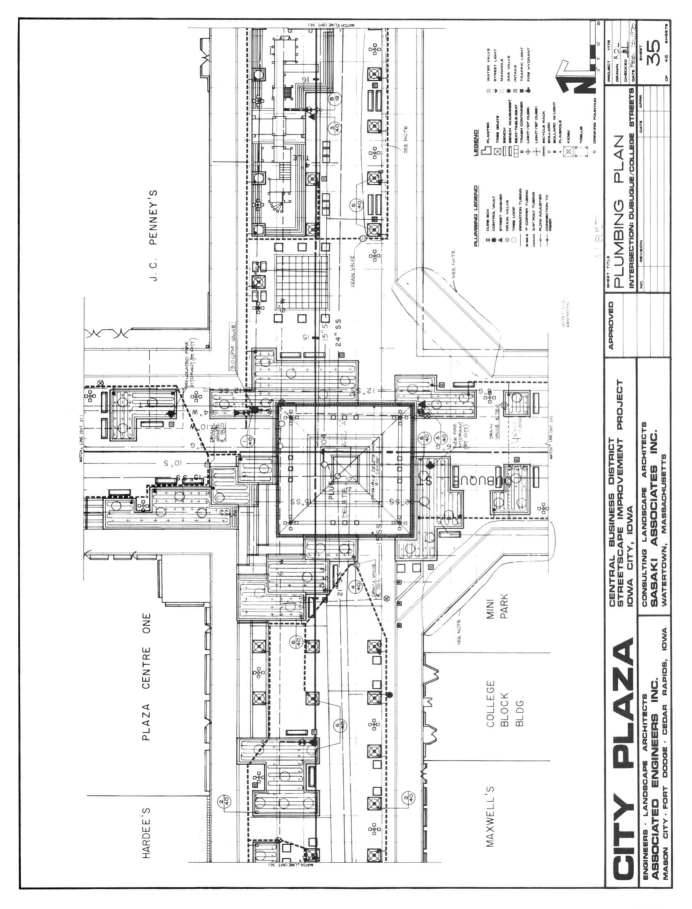

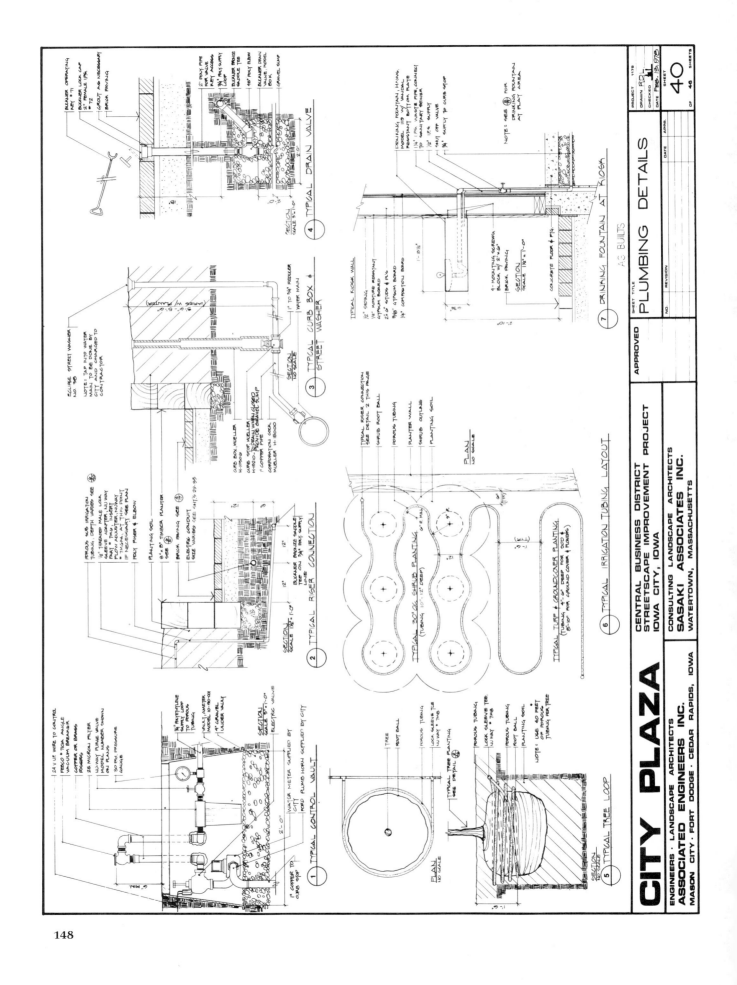

6.14 DESIGN WARRANTY

Aspects of warranty relate to all design work, but there seems to be particular concern about the hidden areas of earthwork and utility system design. A designer finds much confusion in the law and, in general, an imperfect situation with respect to liability.

Essentially a designer is recognized as responsible for due care and skill in carrying out contractual and fiduciary duties. Historically, however, the application of due care and skill doctrines has not necessarily implied a warranty of one's service to a client. Today court findings vacillate between the application of strict warranty to no warranty of a designer's services. Such indecision leaves a designer in limbo concerning potential liability for system failure, material selection, procedures, performance, on-site observation, worker safety, and user health, safety, and welfare.

Several things seem obvious. First, since the 1960s, courts have increasingly applied aspects of strict liability to the design professions—they have viewed the *products* of design services as being subject to strict or limited warranty. For example, in an extreme case a designer might be found negligent for the failure of a system to function properly, even though the designer had applied due care, skill, and judgment common to the design profession. On the other hand, arguments against the application of strict liability to design professionals (or other professionals) are seen as unwarranted and against the public interest. How such ongoing debate and court findings will affect individuals, on a case-by-case basis, is anyone's guess at the moment, but it does seem as though the designer must recognize that responsibility for all aspects of system design, function, and installation must be continuously examined with respect to warranty and liability.

6.15 SAFETY AND LIABILITY

Certainly everyone concerned with construction and site work bears a responsibility for the safety of

Figure 6.10 Typical set of details that describe precise components of a system relative to Figure 6.9. Note the cross-referencing from these details to structural features that bear upon the location or coordination of some underground system's components. The written specifications identify materials, testing, and installation requirements, and the like. (Courtesy Jack Leaman, landscape architect, Ames, Iowa)

all people working on the site or observing it. It is also obvious that no single or simple legal tenet exists that might clearly direct a designer's contract document language or administrative conduct. Consequently the tendency for subsurface system installations to create unsafe situations and the ambiguity of the law in assigning liability can create unsettling conditions.

Since the 1960s, there has been first a move toward and then a move away from the application of strict liability to those who act as an owner's agent. In a sense, the law might be viewed as unclear as to just what part a designer might play in any situation in which a worker was injured on site. Today, however, it is very likely that a degree of liability will be sought on the part of a designer and that the contract documents as well as a designer's conduct will be reviewed in the event that an accident does take place on site.

Assessing all of the legal principles that enter into the placing of liability for worker safety is best left to courts and legal counsel. However, situations, attitudes, language, and designer conduct during the preparation of contract documents and on-site observations bear close examination in light of recent trends:

1. Contract language or verbal directives that tend toward a designer's implicit or explicit control or supervision of a contractor's work procedures or methods may shift liability to a designer. For example, a profusion of language, such as *as directed by . . . under the supervision of . . ., as approved by . . .,* or *with the approval of . . .,* may imply that a designer is not on site to observe but to direct and approve of the contractor's work, in spite of language to the contrary in the general conditions of the contract.

2. Horizontal control (dimensioning) the precise locations(s) of *subsurface* system components, along with detailed control of other excavations, may create designer responsibility for hazardous cave-in conditions.

3. An agent's extreme exercise of *control* over a contractor's work, methods, and procedures may, by implication, remove the contractor from consideration as an independent contractor and form an employee–employer relationship between the contractor and owner (producing complica-

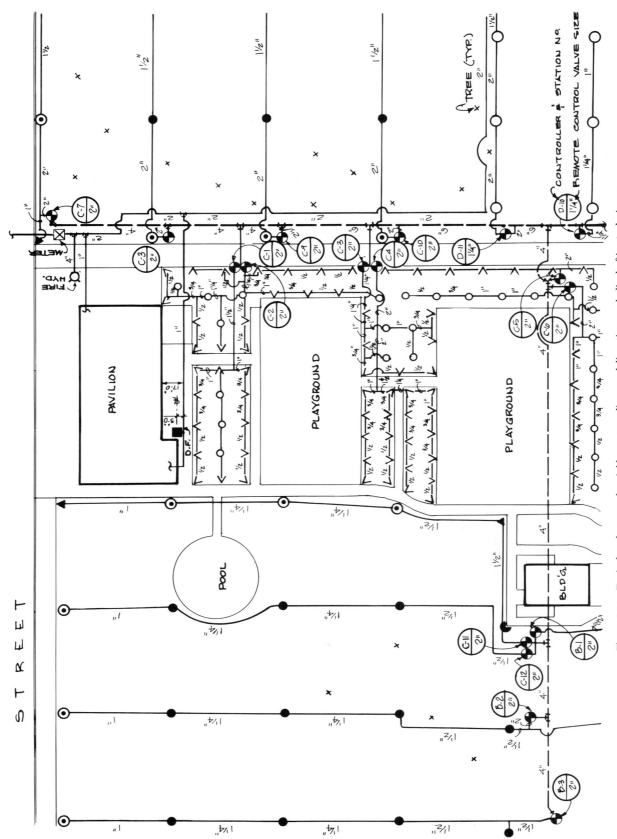

Figure 6.11 Typical underground sprinkler system diagram delineating symbolic drafting techniques and symbols. The system requires large- and small-diameter performance from a variety of sprinkler heads and will operate from automatic clock controllers. Note that existing trees are recognized by routing pipe around these root systems and that potable water sources to buildings and drinking fountains are clearly separated from the irrigation system.

150

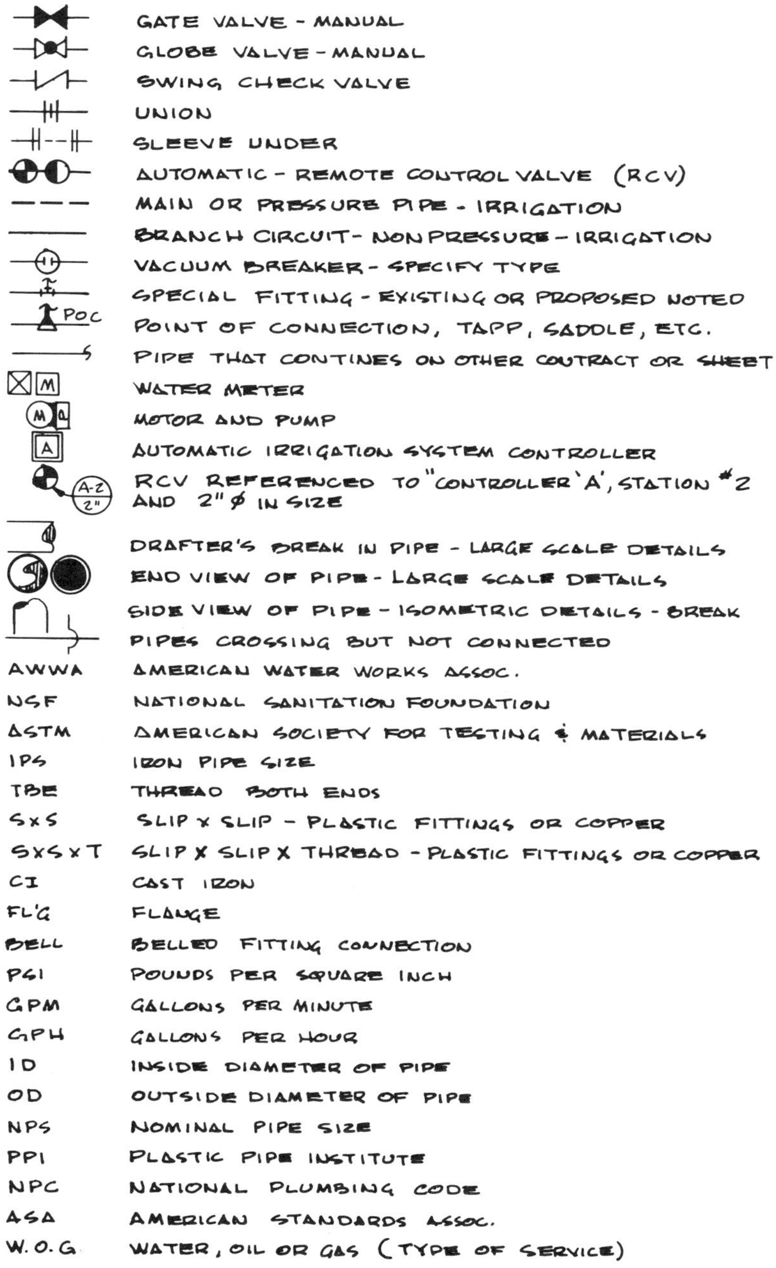

GATE VALVE - MANUAL
GLOBE VALVE - MANUAL
SWING CHECK VALVE
UNION
SLEEVE UNDER
AUTOMATIC - REMOTE CONTROL VALVE (RCV)
MAIN OR PRESSURE PIPE - IRRIGATION
BRANCH CIRCUIT - NON PRESSURE - IRRIGATION
VACUUM BREAKER - SPECIFY TYPE
SPECIAL FITTING - EXISTING OR PROPOSED NOTED
POINT OF CONNECTION, TAPP, SADDLE, ETC.
PIPE THAT CONTINES ON OTHER CONTRACT OR SHEET
WATER METER
MOTOR AND PUMP
AUTOMATIC IRRIGATION SYSTEM CONTROLLER
RCV REFERENCED TO "CONTROLLER 'A', STATION #2 AND 2" Ø IN SIZE
DRAFTER'S BREAK IN PIPE - LARGE SCALE DETAILS
END VIEW OF PIPE - LARGE SCALE DETAILS
SIDE VIEW OF PIPE - ISOMETRIC DETAILS - BREAK
PIPES CROSSING BUT NOT CONNECTED

AWWA	AMERICAN WATER WORKS ASSOC.
NSF	NATIONAL SANITATION FOUNDATION
ASTM	AMERICAN SOCIETY FOR TESTING & MATERIALS
IPS	IRON PIPE SIZE
TBE	THREAD BOTH ENDS
S×S	SLIP × SLIP - PLASTIC FITTINGS OR COPPER
S×S×T	SLIP × SLIP × THREAD - PLASTIC FITTINGS OR COPPER
CI	CAST IRON
FL'G	FLANGE
BELL	BELLED FITTING CONNECTION
PSI	POUNDS PER SQUARE INCH
GPM	GALLONS PER MINUTE
GPH	GALLONS PER HOUR
ID	INSIDE DIAMETER OF PIPE
OD	OUTSIDE DIAMETER OF PIPE
NPS	NOMINAL PIPE SIZE
PPI	PLASTIC PIPE INSTITUTE
NPC	NATIONAL PLUMBING CODE
ASA	AMERICAN STANDARDS ASSOC.
W.O.G	WATER, OIL OR GAS (TYPE OF SERVICE)

Figure 6.12 Typical symbols used in plumbing and irrigation system contract documents. Although typical, symbols may vary among regions and offices.

tions in liability determination and workers' compensation insurance).

4. Designers depend on *indemnity* clauses that are general in nature and not specific to dangerous conditions that may be present during excavations, testing, or other specified operations central to installation of systems.

5. Those having real or apparent knowledge of an *unsafe* situation may fail to correct the condition or have it corrected.

6. During on-site observation, there may be

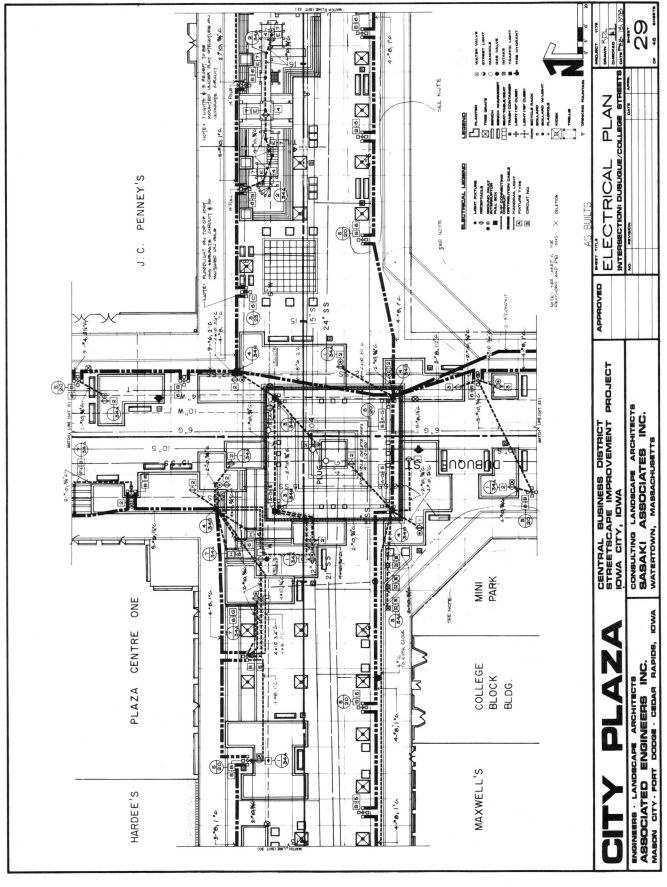

failure to see, detect, discover, or correct *unsafe* conditions.

7. Construction documents fail to make the contractor(s) *solely* responsible for all aspects of on- and off-site safety.

8. Professional documents, agreements, and actions fail to recognize the possibility that an absence of *privity* (an agreement) with a contractor(s) does not necessarily reduce the risk of designer liability for worker safety or injury to third parties.

9. Agents prepare documents that are explicit or implicit in *warranting* the location, depth, material, contents, and the like of a subsurface conduit or feature, and reference to the owner's responsibility for site survey data per the owner—designer agreement and general conditions to the contract.

10. The *agent* fails to advise the owner or the *owner* fails to *stop* a contractor's work whenever unsafe conditions are known to exist on the work site.

The risks associated with the installation of subsurface systems are a leading cause of construction accidents. When system routes and conduits exceed 3 feet in depth, there is a great potential for cave-ins, material falling on workers in excavations, and an assortment of possible encounters with other conduits and obstructions. Wherever contract documents increase the hazards of installing subsurface systems, designers can expect a greater share in the liability.

6.16   CHECK LIST

6.16.1   Single Contractor

The use of a single prime contractor for site work places the contractor in a position to coordinate

activity between finish grade establishment and any excavations for subsurface installation. Contract documents need not sort out or identify the work of several contractors.

6.16.2   Multiple Contractors

Multiple prime contractors will place a burden upon the contract documents to identify the work of each contractor and establish sequential work. The owner or owner's agent is in a position to coordinate, for example, explicit measures of finish among rough grade, finish grade, trenching, compaction, backfill compaction, and system operation relative to each contractor. For instance, the failure of an irrigation system installed by a speciality contractor will affect the acceptance of the site by a planting contractor, the sequence of planting work, potential damage to the irrigation system, and plant guarantee in the event the system fails to function properly.

Possible contract contingencies can be reduced by establishing finish criteria for one contractor and acceptance criteria for another contractor. Completed work of one contractor becomes the existing conditions for another.

6.16.3   Changed Conditions

Contract documents must function as a clear delineation of conditions that will exist at the time a contractor enters a site to begin work. If conditions are not as described in the documents, the contractor may waive such conditions or request a correction. Waivers should be secured in writing and in the form of a complete change order. An agent must take particular care to expose variations in finish grade relative to trench depth and specified minimum and maximum cover over conduits.

SELECTED READINGS AND REFERENCES

Carpenter, Jot D. Editor *Handbook of Landscape Architectural Construction*, Washington, D.C.: American Society of Landscape Architects, 1976.

Foster, Norman *Construction Estimates from Take-off to Bid*, 2nd ed. New York: McGraw-Hill, 1972.

Figure 6.13 Example of an underground electrical system in diagrammatic graphic form. Note the careful delineation of subsurface routing to avoid conflict with planter walls or other subsurface obstructions to installation or, later, maintenance. Note also the importance of a legend to both plan and detail identification. A designer must also control the location of transformers and other above-ground components in order to avoid esthetic problems and inference with pedestrian or emergency movements. See Figure 6.14 for component details. (Courtesy Jack Leaman, landscape architect, Ames, Iowa)

Figure 6.14 (Page 154) Example details of underground electrical system and exterior lighting. See Figure 6.13 for component locations. (Courtesy Jack Leaman, landscape architect, Ames, Iowa)

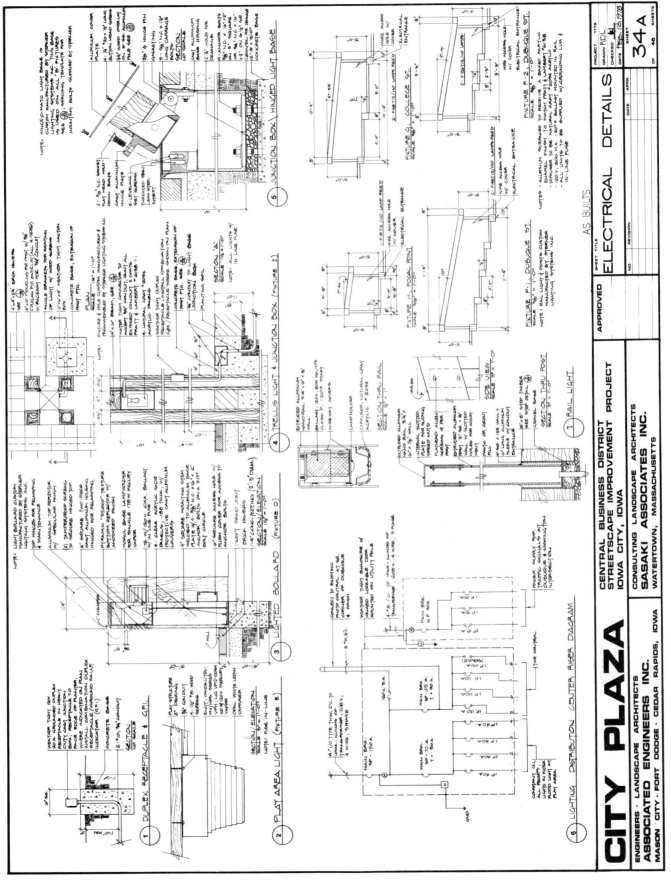

CITY PLAZA

ENGINEERS · LANDSCAPE ARCHITECTS
ASSOCIATED ENGINEERS INC.
MASON CITY · FORT DODGE · CEDAR RAPIDS, IOWA

CENTRAL BUSINESS DISTRICT
STREETSCAPE IMPROVEMENT PROJECT
IOWA CITY, IOWA

CONSULTING LANDSCAPE ARCHITECTS
SASAKI ASSOCIATES INC.
WATERTOWN, MASSACHUSETTS

SHEET TITLE
ELECTRICAL DETAILS

34A

PROJECT 1178
DRAWN RDL
CHECKED
DATE Feb. 12, 1978
OF 46 SHEETS

AS BUILTS

1 RAIL LIGHT

2 PLAY AREA LIGHT (FIXTURE E)

3 LIGHTED BOLLARD (FIXTURE D)

4 TRELLIS LIGHT & JUNCTION BOX (FIXTURE I)

5 JUNCTION BOX / HINGED LIGHT BASE

6 LIGHTING DISTRIBUTION CENTER RISER DIAGRAM

DUPLEX RECEPTACLE & G.F.I.

FIXTURE F-1 DUBUQUE ST.

FIXTURE F-2 DUBUQUE ST.

FIXTURE H FOCAL POINT

FIXTURE G COLLEGE ST.

Ⓙ J	JUNCTION BOX
Ⓛ	LOW VOLTAGE OUTLET WITH RELAY
⊖	SINGLE OUTLET
⊖=	DUPLEX OUTLET
⊜=	TRIPLEX OUTLET
S	SINGLE POLE SWITCH
S₂	DOUBLE POLE SWITCH
S₃	THREE WAY SWITCH
S₄	FOUR WAY SWITCH
Sₖ	KEY OPERATED SWITCH
Sₗ	LOW VOLTAGE SWITCH
Sₗₘ	MASTER LOW VOLTAGE SWITCH
⊖ₛ	SWITCH WITH SINGLE OUTLET
⊖ₛ	SWITCH WITH DOUBLE OUTLET
Sₜ	TIME SWITCH
S_CB	CIRCUIT BREAKER SWITCH
S_WF	WEATHER PROOF FUSED SWITCH OR "SHOCK PROOF"
WP	WEATHER PROOF
RT	RAINTIGHT
DT	DUST-TIGHT
G	GROUND
R	RECESSED
UL	UNDERWRITERS LABORATORY
I.E.S.	ILLUMINATING ENGINEERING SOCIETY
N.E.C.	NATIONAL ELECTRICAL CODE
ANS	AMERICAN NATIONAL STANDARD (SYMBOLS)
▦	PULL BOX
▬	PANEL BOARD (BOX) – RECESSED – FLUSH MOUNT
▬	PANEL BOARD (BOX) – SURFACE MOUNT
▶+	EXTERIOR TELEPHONE
⟶	A TWO WIRE SYSTEM – BRANCH CIRCUIT – RUN FROM FIXTURE TO PANEL BOARD – CONDUIT CONTAINS CIRCUITS #1 AND #2
⫫⟶	SAME AS NOTED ABOVE EXCEPT AS THREE WIRE
⫫⫫⟶	SAME AS NOTED ABOVE EXCEPT AS FOUR WIRE
– – – –	UNDERGROUND CONDUIT OR DIRECT BURIAL WIRE
– – – ⫫	THREE WIRE UNDERGROUND SYSTEM AS ABOVE
Ⓜ	METER (CHECK CONTRACTUAL OBLIGATIONS)
▣	EXTERIOR FIRE ALARM
Ⓐ	LAMP FIXTURE – IDENTIFY FIXTURE TYPE WITHIN LEGEND
○	LAMP POLE
○-Ⓑ	LAMP FIXTURE & POLE
○—	POLE WITH GUY WIRE
▽	TRANSFORMER

Figure 6.15 Typical symbols used for electrical and exterior lighting components in contract documents. Although typical, symbols may vary among regions and offices.

Gray, Donald H., and Andrew T. Leiser *Biotechnical Slope Protection and Erosion Control*, New York: Von Nostrand Reinhold, 1982.

Jewell, Linda "On-Site Sediment Control," *Landscape Architecture*, January 1982, pp. 97–99.

Landphair, Harold C., and Fred Klatt, Jr. *Landscape Architecture Construction*, Amsterdam: Elsevier-North Holland, 1979.

Schroeder, W.L. *Soils in Construction*, 2nd ed. New York: Wiley, 1980.

U.S. Environmental Protection Agency *Processes, Procedures, and Methods to Control Pollution Resulting from All Construction Activity*, EPA 430/9-73-007, October 1973.

Untermann, Richard K. *Principle and Practices of Grading, Drainage and Road Alignment*, Reston, Virginia: Reston, 1978.

Zackrison, harry B., Jr. "Outside Lighting System Design," *Lighting Design Application*, May 1980, pp. 29–37.

Chapter Seven
Plantings

A certain number of plants respond to careful installation, precise moisture control, exact proportions of fertilizer, complete pest control, the absence of disease, and beautiful weather by dying.

John M. Roberts

7.1 A UNIQUE ASPECT OF SITE PLANNING

The installation and care of plantings always add a relatively unique aspect to construction documents and normal contractual arrangements. Plant materials constitute the only living element to be contracted, and thus require an expertise in design and installation that differs appreciably from that for construction. All site work must be directed in a manner that will provide a reasonable horticultural environment during both associated construction activity and the permanent growth pattern of each plant.

Many professionals feel that the contracting of plant procurement and installation is one of the poorest means of achieving excellence in landscape architecture. On the other hand, it is equally awkward to arrange a force account or unit price contract for many site planning projects that require expenditures of public monies or otherwise necessitate close control of expenditures. The preparation of contract documents for plantings may well be the most contrived means of achieving excellence in site design. The following discussion presumes that contracting of planting installation is a necessity rather than a rule.

7.2 THE DRAWING AS SYMBOLIC CONVENTION

A true and precise picture or representation of a plant, its quality, position in space, character, and specific horticultural needs in time, is an impossibility. Such a picture may exist in the designer's imagination but must be translated into reality through the crudest of symbolic language. In a purely *clinical* sense, no one but the designer is really concerned about a plant's ultimate growth or its character after it has been installed by contract method. A contractor must be concerned only with the scope of work outlined within the documents.

A contract drawing's purpose is to indicate the type of plant, its size, and its location on the site. Technical specifications control a plant's quality, and details of installation, care, and guarantee. Any attempt graphically to symbolize a plant's mature size or its botanical character or to describe its purpose is purely informational, not contractual.

7.2.1 Plan Techniques

Several graphic practices prevail in those professions that prepare contract documents for planting design implementation. The choice of graphic technique will relate to local conventions, local training and experience of labor, complexity of the planting design and species, and efficiency of drafting time. Graphics may be loosely categorized into three groups: (1) *embellished,* (2) *graphic symbol,* and (3) *notation.*

1. Some designers tend to mix conceptualization planting design with contract document information. It is often time efficient to add, after an owner's approval, specific plant information to a conceptual drawing in order to create a contract document. Figure 7.1, for example, indicates an *embellished* presentation of what a designer considers to be the ultimate size and symbolic physical character of each plant species. By the addition of

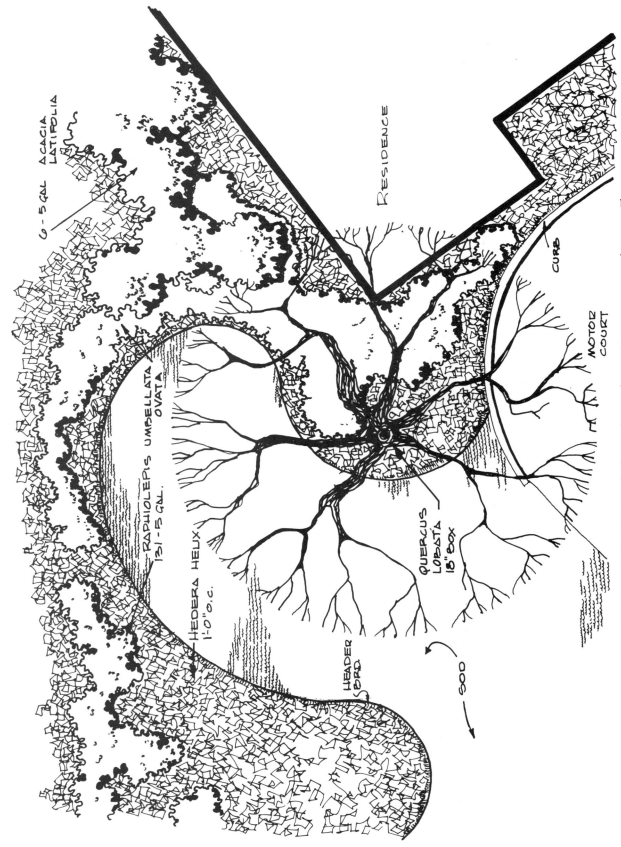

Figure 7.1 Example of overdrawing of a planting design relative to its use as a contract document. The drawing is most suitable for presentation to a client and to show design intent rather than to meet contractual obligations. The addition of plant quantity, spacing, botanical names, and sizes gives the drawing contractual obligations.

158

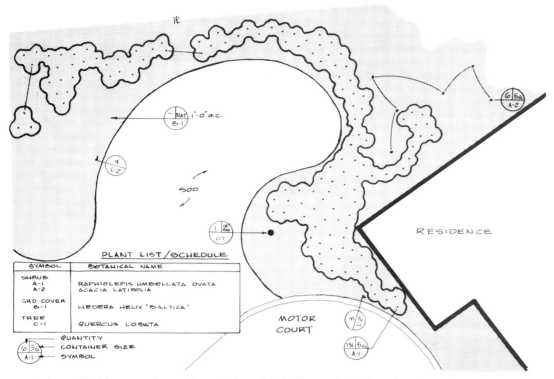

Figure 7.2 The same plan as Figure 7.1 but with the line work significantly reduced for contractual clarification. Symbols and reference symbols are used for precise delineation of a contractor's scope of work.

contractual information, the drawing can become a contract document (as long as such a drawing contains contractual information and the somewhat elaborate graphics do not confuse or obscure that information). However, if a drafter must prepare such a drawing for the *single* purpose of contract implementation, the drawing becomes an example of overdrawing and is a very inefficient use of one's time. All too often, an embellished drawing obscures critical contractual information and is most suitable for presentation to an owner rather than as a contract document. Note that the precise locations of ground covers are obscured by shrub symbols.

2. A second category includes those graphic techniques that completely symbolize the location and contractual information. Figure 7.2 typifies the use of *symbols* that relate to a legend for their contractual meanings. On occasion, each symbol will contain contractual information, such as plant size or spacing. As a completely symbolic language, the plan symbols and legend are *inseparable*. Little attempt is made nor need it be, to show the graphic

relationship of a symbol to the physical character of a plant.

The relationship between the plan symbol and legend may become a source of errors for the contractor, if there is a chance of confusion among symbols. Also, if the legend is physically separated from the drawing, it becomes awkward to identify the various plant species.

3. Contemporary graphics tend to lean in the direction of the third technique, which requires more drafting time than for symbols but is more convenient for the contractor to interpret in the field. Symbols for each plant are simply a dot or X denoting locations, with a *notation* on the side regarding the plant species, size, and, if necessary, the plants' spacing. Figure 7.3 is an example of this technique. Thin, crisp lines symbolically connect each of the plants, which are of the same species and size as noted. Such a drawing is relatively fast to draft, clear in intent, and free of embellishment or other distractions. Note that no attempt is made to identify a plant's ultimate growth habit or size.

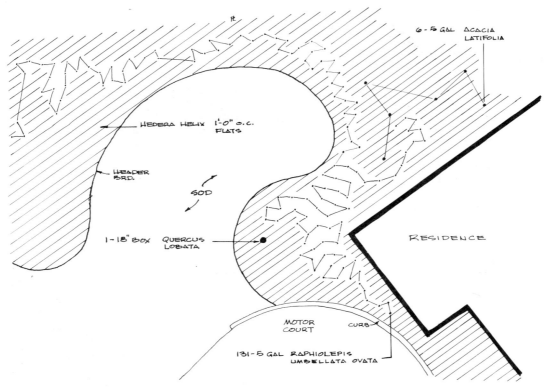

Figure 7.3 The same plan as Figure 7.1 but with the line work significantly reduced for contractual clarification. Symbols are reduced from Figure 7.2 and direct notation reduces the need for a legend or schedule.

7.3   TYPES OF CONTRACTUAL TECHNIQUES

The following discussions present several possible techniques for the contracting of planting installations. Each technique has its own pros and cons but may be suitable for certain owners, local labor, conventions, and complexities of a project. It is always helpful in the development of contract documents if one of the contracting techniques can be selected prior to the preparation of the documents.

7.3.1   Material and Installation

Drawings and specifications delineate a contractor's complete responsibility for the procurement and installation of all plant materials. Selection of this contracting arrangement implies that documents carry complete contractual obligations in precise terms and obligations. A designer's role becomes only "observational" in nature, with the general conditions limiting an agent's role to the approval of plant materials, installation, and maintenance.

7.3.2   Cash Allowance

The preparation of documents and the selection of plant types, quality, sizes, and installation are placed in the hands of a contractor. However, a prime designer may retain the right of approval of these, at contract price adjustment, and, to some extent, of approval of persons executing the work. (See Section 1.8.3.) The use of an allowance arrangement for plantings, unless owner directed, may, at times, constitute an unprofessional transfer of responsibility on the part of those who prepare contract documents or who otherwise agree to represent an owner.

7.3.3   Owner-Supplied Plant Materials

An owner may retain a designer to procure plant materials from several regional sources. Such an arrangement allows both the designer and owner to select only those plants most suitable to the design concept. Selected plants may be purchased directly by the owner. A contract is subsequently let for

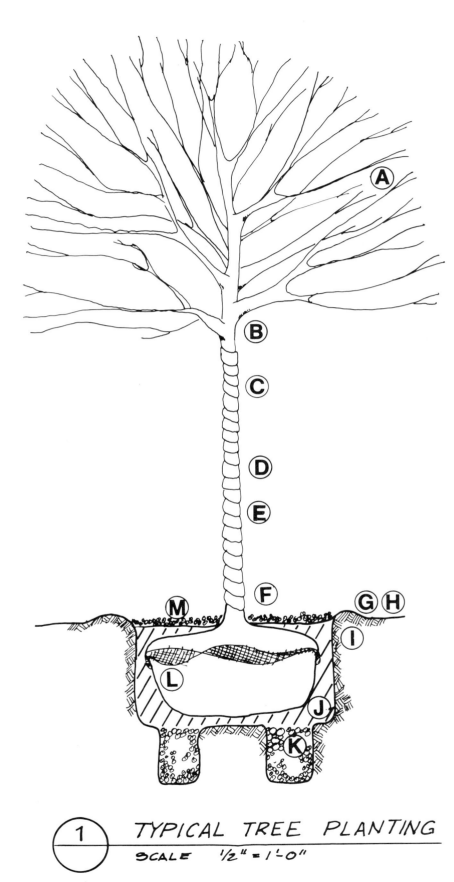

Ⓐ

Ⓑ

Ⓒ

Ⓓ

Ⓔ

Ⓕ

Ⓜ

Ⓖ Ⓗ

Ⓘ

Ⓛ

Ⓙ

Ⓚ

① **TYPICAL TREE PLANTING**
SCALE ½" = 1'-0"

Figure 7.4 Typical section planting detail. A designer must decide, delineate, and specify answers to each of the options when preparing contract documents: (1) requirements for pruning or shaping; (2) the botanical name of each plant and location; (3) protection of the trunk or stem by wrapping or painting; (4) staking or guying requirements as to species and plant size at the time of installation, responsibility and timing removal of staking or guying, safety of pedestrians encountering guy wires; (5) size of plants in terms suitable to local nursery industry and availability; (6) size of pit relative to size of plant; (7) need for and removal of irrigation basins, responsibility for irrigation; (8) location of ground covers and paving to base of tree or shrub; (9) finish grade around plant base and relative to finish grade surrounding plant; (10) characteristics of pit backfill material—import, admixtures, return borrow, remove barrow; (11) special requirements for drainage; (12) protection of root ball against breakage, removal of "burlap," containers, healing in; (13) mulching requirements, depth, material.

161

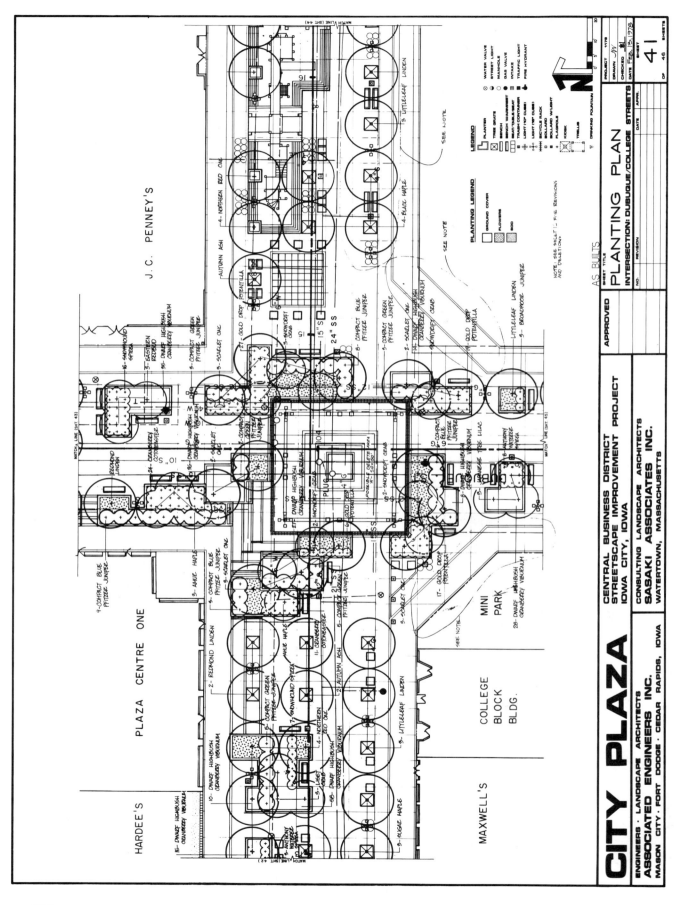

162

installation of the purchased plants. Agreements must identify obligations of delivery and guarantee, as well as planting instruction.

A more complex method might be adopted for use in public or private projects. A designer identifies several selected nurseries or other supply locations. Each particular species of plant is then identified on a bid form submitted to each nursery. The owner purchases plant materials from each bidder and subsequently lets a second contract for the installation of the plants. Again, the bidding documents must clarify responsibilities for delivery to the site, guarantee of the plants, and installation instructions. In some instances, an owner may wish to install plant materials by force account.

Arrangements as noted above allow contract documents to be loosely prepared, if at all. Most of the legal documentation can be accomplished by written technical specification and plants then located on site under a designer's direction. If a drawing is not prepared, there is some danger of over- or underestimating the number of plants needed. The owner must have some means of returning or obtaining extra plants. A contractor will often function best with a unit price agreement for installation. See Section 7.10.3 for further discussion regarding the guaranteeing of plants.

7.3.4 Owner-Grown Plant Materials

An owner may retain a designer to assist in the planning and management of a temporary nursery for the purpose of growing plants for use in large-scale site planning. In most instances, the designer must estimate the future need and species necessary during a project's construction life. Although such an arrangement is not common, it is sometimes necessary in certain economic and supply situations. As with owner-supplied plant materials, an installation contract may be let for the planting operations. Contract documents may or may not be necessary to the success of the venture, depending upon the degree of supervision contemplated by the

Figure 7.5 Typical planting plan for contract document use. Drafting technique is simplified to maximize legibility and precision and underground obstructions delineated for the contractor's information. In this example, common names are used on the drawing for convenience of workers and full botanical information is contained in a plant list incorporated with the written specifications. (Courtesy Jack Leaman, landscape architect, Ames, Iowa)

owner or owner's representative. Guarantees of plant life can become somewhat muddled unless specific directions and obligations are included in the technical specifications.

7.4 BOTANICAL CONVENTIONS

Botanical names are a common language fixed by the International Code of Botanical Nomenclature. A bontanical name (scientific name) usually consists of at least two, and probably three, names for a specific plant. The first name—for example, eucalyptus, quercus, phlox, and juniperus—is a genus of plants with similar physical characteristics. The second name denotes a species, that is, one particular plant within a genus having specific botanical characteristics. For example, *Hedera helix* is a particular plant of the *helix* species within the genus *Hedera* commonly known as "ivy." The genus is always written first and capitalized, the species follows and is seldom capitalized. The full botanical name is underlined or italicized if written in formal text, including technical specifications. When a botanical name is lettered on drawings, capitalization and underlining are commonly ignored.

A third name is necessary whenever a specific variety of cultivar is specified. In 1959, the horticultural term *cultivar* was added to the botanical term *variety* by the International Code of Botanical Nomenclature. However, the two terms are often used interchangeably in all except exacting taxonomic works. Plant varieties named prior to 1959 often retain the conventional varietal nomenclature, for instance, *Hedera helix hahnii*, commonly known as "Hahn's ivy." Present formal convention requires, for example, that *Phlox drummondi 'Sternenzauber,' or petunia 'Rosy Morn,'* be identified by single quotation marks to distinguish these horticultural cultivar from a botanical variety.

The use of only a generic name in contract documents implies a great deal of contractual latitude. Some genera have only a few species, but others have more than 150. However, a document that specifies both genera and species may or may not be specific enough. For example, *Pittosporum tobira* is a species of mock orange that also has a selected variation known as *Pittosporum tobira varigata* (variegated mock orange).

Some plants carry two botanical names. This occurs whenever several taxonomists differ as to a plant's nomenclature or during a period when taxonomy changes a name from that which has common acceptance in the nursery trade. When the condition exists, a designer can avoid problems if both names are specified as being synonymous nomenclature. Many specifications will require one or several published references to guide the settling of disputed nomenclature. L. H. Bailey's *Manual of Cultivated Plants Most Commonly Grown in the Continental United States and Canada* is often used as a referee, as are the *Manual of Cultivated Trees and Shrubs Hardy in North America Exclusive of the Subtropical and Warmer Temperate Regions* by A. Render, and specific local taxonomic publications. The latest edition of *Standardized Plant Names* prepared by the American Joint Committee on Horticultural Nomenclature is most conventionally used.

7.5   PLANT SUBSTITUTIONS

Designers attempt to maintain an idea of plant availability within their region of practice but some plants are just not of the correct size or of acceptable quality at the time scheduled for their installation. Through the years, there have been many attempts to inventory regional availability so that designers, contractors, and the nursery trade can identify species, sizes, locations, and prices. For the most part, however, scattered locations of nurseries, changing patterns of species demand, time necessary for growing stock, and the complexity of maintaining up-to-date records make the solutions less than adequate.

Designers who work at the residential scale can often reserve plants for a client and call for their purchase by a contractor or owner. At the time bids or a price is negotiated, a contractor can include the price of the reserved plant stock in the agreement. Large-scale projects may be forced to depend upon a contractor's ability to locate specific plants in specific sizes among vast numbers of retail and wholesale growers in order to accumulate a sufficient number for the site. Some large projects can estimate quantities of species and have them custom or contract grown several years in advance of their use. Such an operation requires precarious preplanning and the letting of contracts far in advance of actual material installation.

Ordinarily technical specifications' language recognizes that substitution of plant materials is probable. A designer's intent, procedures, and obliga-

tions in processing substitutions are specified. Lump sum contract arrangements are particularly sensitive to language that can protect the owner's interest, the designer's concept, and the contractor's obligations. An owner's interest lies in obtaining what is paid for. A designer is interested in limiting substitutions, approving necessary substitutions, and managing change orders. A contractor should be protected from unreasonable demands and project delays that may result from a designer's rigid adherence to a selected plant species, or failure to respond in a timely manner to requests for substitutions. Particular concern must be given to substitution requests whenever a contractor is functioning under a contract time limit involving penalties, incentives, or liquidated damages.

All parties concerned with site planning must recognize that some plants will simply not be available and arrange their agreement accordingly.

7.6   SPECIES SIZE

A part of species availability concerns the size of plants that may be available. Two methods are commonly used by designers to control and specify the size of plant materials.

The first method is common to small site work involving local plant stock. A designer will simply identify the plant's container size on the drawings. Identification is either by a plant schedule or is lettered directly with each plant located on the site plan. Approval of the plant material quality is usually a professional decision on the part of a designer. A designer's intent is never quite clear but is generally related to what is considered a reasonable relationship between a container's size and a plant's structural frame, health, branching character, leaf vigor, color, and typical growth pattern and form for the species. In many instances, a contractor functions on the basis of a designer's subjective judgment and may have plant stock rejected with reason or unreasonably. It is quite helpful and reasonable for a designer to identify particularly desirable plant characteristics on the plant schedule so that a contractor understands the acceptable plant quality and characteristics. Quite often, a designer will visit the growing site and approve or reject plants before they are shipped.

A second method is common in large site planting projects in which plants are often judged by their average size and character. For example, the publication *American Standard for Nursery Stock* (ANSI Z60.1-1973) by the American Association of Nurserymen can be made a part of the technical specifications by reference. The publication relates the plant ball or container size to caliper, branching characteristics, form, height, rooting, and vigor of typical container or field-grown nursery stock. Acceptance of plant material size is simplified by matching typical characteristics of supplied plant stock to specifications of character. Although such specifications are not totally objective, they do tend to place acceptability within commonly acceptable criteria and dimensional ranges.

7.7 WEATHER AND PLANTING

There is often direct conflict between a designer's obligation to control the horticultural environment during planting operations and the exclusion of a designer from interference with a contractor's methods and procedures. For example, whenever weather conditions produce excessively wet, frozen, or dry soils, an owner, through the agent, may be forced to stop the contractor's procedures or requires alternative methods for planting. Specifications should clearly indicate that any decision that delays a contractor is based upon professional knowledge of horticultural conditions and that delay in the contractor's progress is necessary to protect the owner's interest during and after a contract has run its course. A professional decision to delay a contractor's progress in the interest of plant growth can be a lonely limb upon which to be stranded.

It becomes quite important that such a delaying decision be recorded in order to avoid potential conflicts with the contract's specified time period and material guarantee, to judge the recovery of extra compensation possibly requested by a contractor, to identify the reasons for such a decision, and to accept or reject any alternative methods or procedures offered by the contractor.

In general, weather conditions will become a legal condition expressed ordinarily as an "act of God." No one individual may be held responsible for an act of God, but litigation can be argued over the costs of repairing damage or cleaning up debris. Much of the resulting pain can be mitigated by requiring the contractor to carry insurance to compensate for damage and cleanup.

Lump sum contracts usually consider a contractor solely responsible for the effect of weather conditions and for making all decisions relative to procedures and timing. However, as previously discussed, responsibilities can become somewhat muddled if a designer orders a change in timing, procedures, or methods in the event of or in anticipation of poor horticultural conditions. It can be argued that a contractor's guarantee of plant materials should make a designer's concern for an act of God an overzealous infringement of the contractor's responsibilities. On the other hand, a designer must consider the potential growth of plants over the years as well as the immediate contractual guarantee. However, before a designer recommends that an owner order work to stop because of adverse weather conditions, there must be a firm and defensible horticultural position.

7.8 THE ESTABLISHMENT PERIOD

In general, planting contracts will consist of three stages: installation of material, horticultural control of germination or rooting, and continuous maintenance of the plants. Ordinary construction work or contractual language will not recognize either the owner's or contractor's obligations during each stage, and thus special provisions must be made in the contract documents.

Stage 1 consists of the actual scope of work necessary to *provide* and *install* plant material. At this stage, it is fairly simple to observe and recognize completion of most work and to define a particular day of completion.

Stage 2 is an extremely difficult period for both contractor and designer. This stage, the *establishment* period, begins on the same day that stage 1 is completed and ends the day that *maintenance* begins. Stage 2 may not be necessary on every project, however.

An establishment period is usually necessary to determine a contractor's proficiency and to avoid an owner's assumption of risks associated with horticultural conditions affecting rooting, germination, or transplant shock. Such a period is associated with a scope of contractual work involving the skill of a contractor. If, for example, a contractor has executed the scope of installation work and has skillfully nurtured seeds during their germination period, the establishment stage has served an owner well. On the other hand, if it is presumed that a

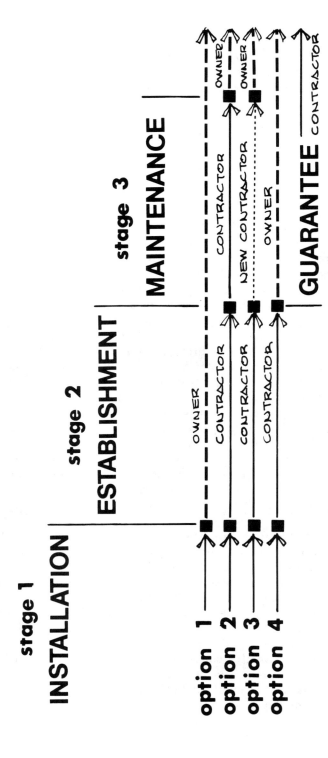

Figure 7.6 Diagrammatic representation of optional contractual relationships among stages of *installation, establishment, maintenance,* and *guaranteeing of plant materials.*

contractor has executed a seeding operation with due care and skill whereas the lay owner attempted to achieve germination but failed, a great deal of argument will ensue. Everyone is presumed guilty of poor performance, the weather will come in for its share of accusation, and no one will be comfortable with the situation.

An establishment period is necessary to a designer in order to measure a contractor's compliance with seeding, stolonizing, transplanting, and other operations that are not visually obvious as having been completed or correctly accomplished. A designer who periodically observes planting operations can approve payments requests and honor contractual obligations only after germination of growth can be observed and evaluated. If a designer is responsible for periodic observation of a contractor's work, an establishment period is necessary to measure contractual performance and should not be an owner's option.

The duration of an establishment period may pose contradictions for all parties. A calendar period is preferrable as it has a precise beginning and end, but, with uncooperative weather, its selection may prove unfair to the owner. For example, if a contractor performs with due care and skill but without success of germination, the owner will not see any plants. The owner is prone to blame the contractor and designer while the weather might have been the culprit. Contractually, the owner might end up paying for the same work two or three times before germination is successful. The contractor has been relieved of protracted operations and has received extra compensation by change order but the owner has borne the full force of risk. An owner's risk is somewhat limited in regions and seasons of relatively stable weather patterns and with daily on-site observations of a contractor's efforts.

On the other hand, a common procedure is to allow the establishment period to remain "open ended" with some sort of subjective or objective means to measure success. The highest risk to any contractor is provided by the statement, "The establishment period will end when approved by the landscape architect." Only those contractors with the innate ability to read a designer's mind will be able correctly to estimate the length of such an establishment period. Risk can be lessened if contract documents can indicate objective measurements—for example, height of grass blades, percent of soil surface with observed germination, observation of plant's turbidity and nonwilted con-

dition, observation of stolon rooting and blade growth. Although they are somewhat subjective, these are attempts to describe success or failure in terms that are common to botanical convention and ordinarily recognized as having a degree of fairness. Under these conditions, a contractor must request an on-site observation of the established plant growth and the designer must notify the owner of success or failure of the contractor's work.

A designer must take care not to describe a contractor's operations or procedures during an establishment period. Certainly a contractor's general duties may be outlined, such as irrigation, top dressing, and rolling, but any attempt to control the degree of care may cloud a contractor's obligations and control of the work.

A third method of defining an establishment period combines the best elements of the other two. For example, a minimum number of calendar days is specified during which a contractor will perform those duties necessary and common to establishing plant growth. At the end of this period, usually two to three weeks, a contractor calls for an inspection. The owner's agent observes and measures growth against a set of criteria, such as height and percent of coverage, and either accepts or rejects the contractor's work. If the work is rejected, an extension of time is agreed upon and another observation scheduled. In some offices, the cost of a second or third observation is borne by the contractor.

At the completion of an establishment period, three options remain for the owner. An owner's maintenance force may assume control the same day that the establishment period ends. A second contractor may enter the site to begin maintenance under an agreement separate and quite distinct from the initial installation and establishment agreement. An installation and establishment contractor might continue under the original prime contract and assume maintenance duties immediately following completion of the establishment period. Selection of one of these options must be made by the owner during the design development stage of the agent's work and specifically outlined in the documents. In addition, the contract documents must explicitly state the beginning of any guarantee periods to be borne by any of the contractors and obligations of all parties relative to replacement of plants that may die during or after the various stages of the contract agreement.

Although guarantees are discussed in more detail later, it must be noted here that a contractor's guarantee will be tied directly to the period of establishment and the maintenance period. It is critical that an owner's maintenance not confuse guarantees or warranties by overlapping the obligations of a contractor.

Both the establishment period and the maintenance period are exceptionally risky ventures for any contractor. In most instances, it is akin to placing one's economic faith in local weather patterns and equally unpredictable horticultural environments. Some contractual risk might be shared by the owner and contractor under the following circumstances.

1. Large construction sites or open spaces are zoned so that a contractor can finish installation, enter an establishment period, and, if required, begin a maintenance period while simultaneously completing the next zone of work.
2. Specifications clearly define the conditions of acceptance and rejection of the various periods of obligation.
3. A contractor is required to replace all plants that die or are otherwise unacceptable during the establishment period but is allowed to share costs for plant replacement with the owner during the maintenance period.
4. Rather than bid maintenance as a part of a lump sum planting contract, the contractor can function by unit prices as necessary to meet whatever unknown conditions may occur during the maintenance period. Perhaps the owner can supply such materials as fertilizers and insecticides and the contractor apply them in an attempt to share risks without lowering maintenance standards (see Section 7.9).

7.9   THE MAINTENANCE PERIOD

Among those contractual obligations of site work, maintenance of plantings is quite high on the list of concerns. While most public agencies and many private clients assume that bidding under a set of specifications can fulfill their wishes, the fact remains that very little is accomplished outside of the good horticultural sense of a contractor. Ordinary construction and installation actions can be adequately prescribed and described by contract documents, but maintenance of plant materials

seems constantly to elude the best of contractual intentions.

The most that contractual maintenance specifications seem able to do is to develop a framework within which a contractor may function, that is, to present guidelines as to what the contractor should accomplish. Very little can be achieved by words regarding the precise measurement of performance. In most instances, a professional opinion is necessary to evaluate a contractor's work relative to the horticultural environment of the site, to judge reasonable attention to the horticultural environment, and to decide whether or not a contractor has provided due care and skill in a given set of circumstances.

7.9.1   CONTRACTUAL SITUATIONS

Three contractual conditions will usually exist toward the end of a planting contract. First, the contractor who has been responsible for installing plantings may include short- or long-term maintenance as a part of the original agreement. An owner may require plant material maintenance *before* accepting the planting contract. This ordinarily will require the designer's attention, observations, and possibly consultation during the maintenance period and will have a direct impact on plant material guarantees. The second possible contract method involves the owner's request and acceptance of proposals and costs, unit or lump sums, from a variety of contractors. If proposals are requested, contractors will submit descriptions of their scope of services, materials, and the like, along with prices. Normally an owner supplies only the barest of information as to requirements. A third method might involve the preparation of fully documented contract and technical specifications to be reviewed and bid by a variety of contractors. The "winning" contractor then functions under a set of directions per the contract documents. In all three instances, it is possible for the owner to assume some specified duties under force account or employee procedures. The second method obviously requires much expertise in sorting out least cost for excellence in maintenance. Methods 1 and 3 require a good deal of expertise in preparing contract documents.

A complete discussion of plant material maintenance is beyond the scope of this text. However, when specific circumstances require specification development, the following check list can serve as a content guide.

1. *Acceptance.* Presume that either a new contractor or the planting contractor will accept or reject conditions existing on the day plant maintenance commences. For example, specify a site visit by the bidder or contractor, owner, and owner's agent prior to or on the date maintenance begins. It must be clear that a maintenance contractor or bidder will assume responsibility for conditions that exist at the inspection time. If a new contractor is involved, allow for development of a *punch list* of nonresponsibility before beginning maintenance. Either the original planting contractor or the owner must assume corrections to problem conditions or the new contractor may negotiate corrections *before* the maintenance period begins. Check for:

a. Complete and adequate irrigation system operation;

b. Adequacy and completeness of any previous contract work;

c. Need for additions or corrections outside the scope of the maintenance specifications;

d. Responsibilities for additions or corrections to the site.

2. *Furnishing of equipment.* Delineate contractor and/or owner responsibility for supplying necessary equipment, tools, operators, and the like, for maintenance.

3. *Performance.* Describe, in performance terms, what it is the contractor will be responsible to do. Among possible terms are aid, alter, arrange, ask, change, clean, close, collect, combine, communicate, complete, conclude, contribute, cooperate, correct, cover, cut, decide, design, discuss, distribute, empty, explain, extend, fasten, fill, find, finish, fit, fix, follow, formulate, guide, hang, help, hold, infer, itemize, join, label, lay, locate, make, map, mark, match, measure, mend, modify, mow, name, note, omit, open, order, participate, pay, perform, permit, pick, place, plan, position, predict, present, produce, propose, provide, prune, raise, rearrange, recombine, reconstruct, record, reorder, repeat, replace, restore, return, revise, ride, rip, save, select, send, separate, serve, share, sharpen, shovel, shut, simplify, sort, sow, spread, stake, start, stock, store, substitute, suggest, supply, support, take, test, twist, uncover, use, vary, wash, and work.

4. *Nonresponsibility.* It is critical to clarify exterior maintenance operations that are *not* a part of the planting maintenance contract. Those on the following list may or may not fall under some exterior planting maintenance and management contracts.

a. Repair, including cost and labor, of irrigation system equipment;

b. Swimming pool and pool deck maintenance;

c. Snow removal and clearing or removal of ice from walks;

d. Parking lot, driveway, or road maintenance, sweeping, and striping;

e. Owner acceptance and negotiation of "acts of God";

f. Replacement of dead or dying plants as a result of disease, theft, vandalism, and the like;

g. Testing of soil conditions, recommendations of fertilizers;

h. Performance of tree surgery required because of age or weather conditions;

i. Staking and guying or their removal during maintenance;

j. Seasonal replacement of annual and perennial "color" plants;

k. Removal of debris and trash off site;

l. Control of weeds in areas of paving, such as parking lots, driveways, and walks;

m. Maintenance of storm drainage systems, inlets, discharge points, headwalls;

n. Special management of turf, for example, verticutting, thatch removal, aeration, greens mowing;

o. Use of special equipment and procedures for deep watering of plants during periods of severe drought or turf aerations;

p. Repair of on- and off-site erosion conditions not caused by maintenance procedures;

q. Repair or replacement of owner-supplied equipment or tools;

r. Exterior maintenance of building surfaces;

s. Replacement of play yard equipment, surfaces, and fencing, or liability for play equipment functions;

t. Painting, staining, or reconditioning of exterior fences, screens, or walls;

u. Repair, painting, and replacement of luminaires, or replacement of lamps for exterior lighting;

v. Care of special turf areas, such as golf greens and tees;

w. Care of special sports facilities, including ball diamond infields, tennis courts, and picnic areas;

x. Care and maintenance of restrooms;

y. Care and maintenance of pedestrian bridges;

z. Repair, replacement, and care of loose paving forming paths or trails;

aa. Management, repair, replacement, seasonal conditioning, and cleaning of ornamental fountains and pools;

bb. Maintenance of exterior drinking fountains;

cc. Cleaning, weeding, and maintenance of permanent silt traps or removal of temporary facilities;

dd. Care of garden ornaments, sculpture, benches;

ee. Seasonal overseeding of cool-season grasses;

ff. Care of artificial turf;

5. Tree, shrub, and plant bed care. Ordinarily activities such as pruning, clipping, edging, weeding, watering, removing spent flowers, thinning, removing fronds and suckers, seasonal care, preventing underground runners, mulching, fertilizing, controlling pests and weeds, and replacing plants are a part of exterior work.

6. Interior plantings. Interior planting activities include washing and polishing foliage, replacing mulch, planting bulbs and tubers, removing excess water from containers or planters, "misting" or otherwise controlling humidity, and protecting carpets or other surfaces during all maintenance procedures.

7. Mowing. Although specifying the number of mowings per week is a simple procedure, it is better, because of weather variation, to direct acceptable minimum and maximum grass heights to be maintained. Cutting heights must recognize acceptable culture for each specific grass type. It is always best for an owner to request an expert review of cutting heights during the maintenance period, because much turf health is tied to this operation. Mowing height may vary seasonally and daily for weed and disease control.

8. Edging. Usually lawn edges at beds and paving are cut at every other mowing. However, less frequent edging will save money if a less than manicured appearance is acceptable.

9. Aeration. Aeration is a difficult operation to predict, but most turf will require it at least twice a year. The equipment is very specialized, and is available as core or spike. The owner may wish to use this operation under an auxiliary agreement when requested.

10. Verticutting. This may or may not be necessary with certain types of grasses and environ-

mental conditions. The equipment is very specialized. When this operation is requested, the owner may wish to use it under an auxillary agreement that also covers off- or on-site disposal of thatch material.

11. Fertilizing. This is an extensive subject that requires great care in determining need, materials, and timing. The specifier is faced with at least three positions in contract documents. First is the specification in performance terms of the pounds of *actual* nitrogen, P_2O_5, potash, minerals, or the like, that must be applied by a contractor each season or year. Such a technique allows the contractor latitude in selecting fertilizer materials, frequency and method of application, and other procedures. A contractor can easily verify the *actual* amount of nutrients applied. A specifier will no doubt wish to determine *actual* quantities of nutrients required by expert testing and opinion for a specific site or portions of the site.

A second method prescribes both the type and quantity of fertilizer materials to be applied at certain times of the year. Professional judgment and local experience must substitute for any specialized tests. The single advantage of this method is its simple approach to contractual obligations.

The third method calls for a cooperative management program between owner and contractor. For example, an owner hires an independent laboratory to provide soil tests of horticultural conditions and make recommendations. The owner and contractor then negotiate for material and application costs. If unit prices for applications are a part of the agreement, the owner might provide materials to the contractor and the application costs will be covered by the established agreement.

12. Fertilizing—interior. This is similar to the fertilizing principles discussed above, except that many special conditions exist. For example, smells of fertilizers are generally a problem indoors; diluted liquid fertilizers possibly are best, and tablet-type fertilizers might be used to reduce smells. The timing of the activity must match the use of the space by humans. Many plants depend upon their foliage for an esthetic appearance; usually fertilizers are specialized and heavily mineralized. Particular discretion and care should be exercised for plants in containers.

13. Irrigation. Irrigation is a subject requiring skill and knowledge on the contractor's part with little, if any, true direction given by specifications. Certainly modern instruments, such as tensiometers, and automatic controls can be objective measurements of performance, but, on a day-to-day basis, they cannot completely substitute for a knowledgeable and concerned person. In general, a specifier must depend upon intent: observable plant growth; consistent moisture available to plants; use but not waste of water; operations not disruptive to human uses of space; schedules related to disease, pest, and weed control; and maintenance of a balanced equal relationship among soil particles, water film on soil particles, and oxygen in the soil.

The specifier must precisely control a contractor's responsibility for maintenance of any irrigtion system or the provision for any irrigation without a system. For example, a sophisticated underground sprinkler system will require constant attention and minor or major repairs. If no system is available, will the contractor be required to use water hoses, buckets, or other means during portions of the year or will the entire site depend on natural rainfall? If a sprinkler system is used, will the contractor be responsible for repairs under some unit prices, or be forced to accept all risks of system repair?

14. *Irrigation—interior.* Similar conditions prevail in interior areas to those noted above. However, the control of moisture and oxygen in the soil mass becomes extremely critical in planters, containers, and closed heated and ventilated areas. Almost daily checking of moisture conditions is necessary, surfaces must be protected from spilled water, overwatering must be pumped out of planters, water containing fertilizers may leave white salts on leaf or planter surfaces, and care must be exercised to ensure water penetration of root balls or plants that remain in their original nursery containers.

15. *Pest and disease control.* This is a complex issue subject to local licensing requirements for advice as well as control measures. The contractor must note all applicable laws and regulations. Under extreme conditions, the contractor may be required to store and use tools and equipment on site to prevent the possible transportation of diseases and pests from other sites.

The subject keys directly to cultural activities associated with mowing, pruning, irrigation, and similar procedures. The owner may wish to share risks for materials and labor or place the entire risk upon the contractor. All fumigation methodology must be controlled.

16. *Pest control—interior.* In many ways, pest control is much more complex in interior areas

than on exterior sites. Sprays, dusts, and the like are safety hazards and also may produce odors that are injurious to clothes or are otherwise objectionable. Dilute control solutions must be used. Constant checking is necessary to detect minute insects and diseases that are very aggressive in climate-controlled zones. Aerosol use should be prohibited, as well as any fogging device with airborne particles. Otherwise specifications are similar to exterior controls but have a greater emphasis on cleaning and sterilization of maintenance tools.

17. *Work force.* Workers are controlled by local regulations, and should be employees of an independent contractor. Minimal workers' compensation insurance against bodily injury, and property damage and aggregate insurance should be provided. Such insurance needs should be discussed with an expert. The work force should be restricted to those trained in and capable of operating equipment and providing horticultural expertise. Constant supervision by the contractor or contractor's agent should be provided.

18. *Solid waste and debris.* All waste and debris produced by the contractor's operations should be removed daily and the area swept. It should be decided whether the contractor is to pick up and remove waste and debris by others from all areas, including parking lots, drives, recreation areas, and from trash containers, and when.

19. *Weed control.* Control of weeds is normal to plant maintenance activity. It may involve local controls over chemical usage as with pest control. Use of oil-based chemicals should be limited to nonpaved zones; herbicides are best for all areas but oil-based types may be less expensive. Questions to be resolved include the removal and disposal of dead weeds, whether the contractor is to control weeds in cracks through walks, parking lots, and driveways, and how large weeds will be allowed to grow before control. Specifically the use of fumigation and both permanent and temporary soil sterilization practices should be controlled.

20. *Extras.* Those activities that are to be considered outside the contractor's scope of service but may be required seasonally or occasionally should be identified and defined. The cost or cost range should be ascertained prior to the agreement's execution.

21. *Storage.* Control of the contractor's storage of equipment, tools, materials, and equipment repair activities on site is important. The owner may or may not wish the site to be used by the contractor.

Particular care must be taken with storage of the contractor's chemicals or materials on site relative to the owner's liability for the health and safety of those occupying the site.

22. Burning. Is burning of debris, solid waste, leaves, and the like, allowed on site?

23. Seasonal work. Any seasonal work should be identified, such as leaf raking and disposal, wrapping of tree trunks, cold weather protection, mulching, pruning, fruit removal, transplanting, overseeding, and erosion control planting.

24. As-built drawings. The contractor must have access to any drawings of the irrigation system as installed with emphasis on the locations of pipes, sprinkler heads, controls, automatic controllers, pipe materials and sizes, metering, pressures, guarantees, warranties, pump performance curves, and, if possible, discussions with the system designer.

25. As-is drawings. The contractor should be made responsible for recording all revisions, additions, or corrections to the irrigation system on a set of owner-provided documents. The owner should retain possession of such as-is documents.

26. Continuity. Any additions, corrections, or deletions regarding the plantings, irrigation system, or site work should be made in accordance with any previous contract document details and specifications made available to the maintenance contractor by the owner. Any deviations are to be made only with the owner's approval.

27. Guarantees and warranties. A maintenance contract may span all or a portion of a previous contractor's guarantee or warranty period. If this is so, the owner must avoid conflicts among contractors and possible loss of guarantees or warranties.

One method of reducing potential conflict requires the contractor under guarantee or warranty to provide two to four site inspections per year as a part of the planting contract. The planting contractor will supply the owner and/or agent and the maintenance contractor with copies of each inspection report. If the planting contractor finds any indication of improper procedures that may affect the guarantee or warranty, these are reported. Failure to report may nullify the planting contractor's defense that, for example, improper maintenance caused or may have resulted in the death or poor growth of guaranteed work and warranted plant materials. Note that such procedures begin in the contract documents covering plant material installation. It is recommended that such proce-

dures be used even though the owner executes maintenance or has it executed.

It is critically important that the exact nature of guarantees and warranties be related to the quality of maintenance received by living plants during a guarantee or warranty period, regardless of who provides maintenance.

28. Bonds. The owner should investigate the feasibility and costs associated with requiring the contractor to secure a performance-type bond to run for the duration of the maintenance contract.

29. Agent observations and reporting. A designer's ultimate involvement in preparing maintenance specifications will run with the original owner–designer agreement describing services or as an extraordinary agreement to be executed at a later date.

30. Reference. Because of the interrelationships among proper horticultural activities, plant materials, weather variation, soils, water quality, and horticultural skills, it may be advisable to make certain horticultural publications a part of contract specifications by reference. For example, publications regarding local turf grass management might best be referenced rather than trying to anticipate all variables within the limited confines of a contract.

7.10 PLANT GUARANTEE

A guarantee of plant material differs appreciably from that of ordinary construction and utility systems. The major problem is to objectively measure whether or not a contractor is or is not responsible for the life or death of a plant. Quite often, it is impossible, at least visually, to determine whether a plant is alive or dead during certain seasons of the year, and depending upon its characteristics.

The common guarantee period runs for one year following completion of an establishment period. During this period, a contractor is directed to replace any materials that are "dead," "dying," "fail to grow," or "fail to produce hair roots," and a sundry list of abstract words that attempt to deplict intent. The fact remains, unfortunately, that there is no clear or objective procedure to identify those plants that require replacement as a result of a contractor's failure to live up to the spirit of a contractual obligation.

The language of most guarantees assumes that a plant's death or poor condition is prima facia

evidence of poor installation procedures. An owner
faults the designer and contractor. A designer faults
the contractor and, in some instances, the owner.
On occasion, none of the people involved are guilty
of anything. The art of postmortem on a dead plant
is not advanced enough to ascertain the exact cause
of death.

To study the problem of guarantees, let us first
examine the interests of the parties involved.

7.10.1 The Owner

An owner's interest is, of course, in obtaining exactly
the quantity and quality of materials paid for. As far
as most owners are concerned, they do not wish to
be involved in a scientific or botanical explanation of
why a plant appears to be dead but really is not. If
the plant appears to be dead, then it is dead, and
must be replaced under the contractor's guarantee.

Behind the scene, an owner's negligence in
maintenance may have been totally responsible for a
plant's demise. The owner may have selected the
plant at a nursery and insisted upon its incorporation in the contractor's work, in spite of the designer's or contractor's protest. A plant may have
been grown by the owner. A plant may have been
purchased by the owner from a grower with whom
the contractor had no association.

7.10.2 The Nursery

Two types of plant nurseries generally exist as
separate or combined functions and operations. A
retail nursery will grow or purchase plants for resale.
Quite often, such a nursery offers purchase and
professional discounts when plants are sold to
landscape contractors. A second type of nursery
functions as a grower of plants for wholesale distribution to retail nurseries, and possibly as a source
of plant materials for landscape contractors who
require large quantities of plants. The guarantee of
plant life may be traced to the plant grower but there
is no contractual connection among the nursery,
owner, and contractor. If a plant dies, a nursery's
involvement with a replacement must be negotiated
between the contractor and the nursery. Neither the
owner, the designer, nor the contractual specifications may implicate a nursery in guarantee issues
(except, perhaps, within the guidelines of any court
litigation as might ordinarily involve a supplier of
materials for a construction contract).

7.10.3 The Designer

As an owner's agent, a designer becomes involved
with a guarantee as its author and administrator.
Unfortunately, whenever a plant dies, a designer
must make a decision as to the reasons and assign
responsibility. In some instances, a designer may
find that an owner's negligence or overzealous
treatment of a plant was probably the cause of death.
On the other hand, the contractor, as the installer, is
usually viewed as guilty until proved innocent. In
some circumstances, a designer may admit, with
commendable honesty, to selecting and specifying
the wrong plant for the site.

7.10.4 The Contractor

The contractor is, of course, the one person who
must answer a guarantee. At issue with a plant
materials guarantee is to what extent the contractor
feels a risk is assigned through the contract documents. Specification language that places degrees of
risk for replacements upon a contractor no doubt
will cost the owner in contigencies for such a risk.
Specification language that is unclear as to the
responsibilities of any of the parties must be interpreted as laying the total responsibility on the
contractor.

An experienced contractor realizes that it is next
to impossible to predict the horticultural conditions
at the time of planting, the reaction of a site's soil
chemistry to the plants, and the arbitrary way in
which plants seemingly die in spite of all efforts. The
only means of dealing with guarantee conditions is
either to estimate financial risk as a contingency to
performing the planting work or to submit both the
designer and owner to delays until someone wears
down.

7.10.5 Conditions of Guarantee

In developing language of a guarantee, the following
concepts require definition and intent.

1. *Time.* What is the duradtion of a guarantee?
In many respects, a guarantee's duration is tied
directly to the regional climate and the type of plant
materials installed. Cold climates combined with
deciduous plant materials require at least one
growing season before a plant's condition can be
assessed. If the material is planted as bare root
material, the guarantee must carry through one

leaf-out for assessment. Mild climates and evergreen materials can generally allow determination of a plant's general well-being after only a few months. However, coniferous plants will require at least a six months' trial or the advancement of fresh growth to assess condition.

2. Species. Will the guarantee be a blanket one for all species or will different times and conditions be placed on various species and species characteristics? Very often, a contractor's risks can be reduced if each species or type of plant is treated individually and judged on criteria befitting the plant's growth habits.

3. Procedures. Will the contractor be notified by the owner or designer of the plant's need for replacement or will decisions be made as a team? Notification by the owner or designer may be quite dictatorial in interpretation. A team effort can often resolve conflicting opinions on site with an increase in everyone's understanding.

4. Container- or field-grown materials. Will the guarantee distinguish among the various types of contractor responsibilities for transporting and growing each plant? Container-grown stock often carries a reduced time of guarantee relative to field-grown stock. The risk of guaranteeing container-grown material is often less than for material that must be dug and transported as bare root, balled and burlapped, or boxed.

5. Criteria. Is there some means of identifying and describing exactly or functionally the basis upon which replacement plants will be defined? Can a percentage of growth, leaves, new growth, wilting, die-back, mechanical injury, or color, or some other method be found to determine when a plant must be replaced? That is, is there any deviation from installed perfection that may be considered acceptable?

6. Size. Is there a relationship between the time or type of guarantee and the size of a plant when installed? Small plants can be quickly identified as to condition whereas larger plants generally require more time to recover from transplanting shock.

7. Personal responsibilities. Is the contractor responsible, for instance, for making specific visits to the site to inspect the owner's maintenance and advising the designer of any conditions bearing upon the guarantee, or for completely maintaining the plants during the guarantee period, or advising the owner of reasonable maintenance practices? Is an owner able reasonably to maintain the plants during a maintenance period and note that in the agreement with the contractor? Is the designer to advise the owner concerning maintenance conditions during the guarantee period or to make periodic site visits to determine maintenance procedures and conditions?

8. Consultants. Is provision made for the expense and duties of any individual retained to render an unbiased opinion as to conditions bearing on the guarantee?

9. Default. Is there any provision for a contractor's default in the replacement guarantee? Special conditions to the contract may allow the retention of monies in an account to be released upon completion of the guarantee responsibilities (including accrued interest). A contractor may be bonded as a guarantee of the owner's expenses in the event of default.

10. Act of God. Does the guarantee provide for a contractor's release from the guarantee in the event of acts of God that damage, break, kill, stunt, defoliate, disease, burn, or otherwise harm plants under the guarantee? A contractor must be held responsible for workmanship and skill quality, but not for circumstances beyond control.

11. Vandalism and theft. Is the contractor to be held responsible for acts of vandalism? It is not unusual for a guarantee to require replacements for vandalism and theft but the owner must carefully assess the site's location and potential risk to be taken by the contractor and owner.

12. Unit prices. Is it feasible for a bid to request unit prices for the replacement of certain plants that may be judged to be in need of replacement but are not the responsibility of the contractor? This is an unusual request but one worth considering in light of the owner's and contractor's interest in the completion and success of a particular planting.

13. Shared risk. Does the guarantee provide for a sharing of risk between the owner and contractor? For example, in the event of plant failure, will the owner provide a new plant and the contractor replant?

14. Postguarantee. Are replacement plants to carry a guarantee or does replacement end the contractor's obligation and complete the agreement? Perhaps this is a logical protection of an owner's interest, but the process is difficult to record and can be time consuming.

15. Designer's agreement. Does the designer's agreement with an owner require the time necessary to administer, inspect, and record the specified contractor's guarantee period?

16. ***Period of guarantee.***   Does the period of guarantee include or exclude any period during which a contractor must provide maintenance of the plants? The answer to this lies in defining when maintenance begins and ends with respect to the guarantee period.

7.11   PROCEDURAL SPECIFICATIONS AND GUARANTEE

It is common for today's professional to prepare planting documents in traditional procedural language. As procedural language, specifications dictate both materials and their installation. Such a tradition has probably resulted from a designer's movement away from personal contact with planting work and exemplifies an unwillingness to lose control of the manner of installation. The procedure is aggravated by a designer's inability to develop performance-type specifications. There is simply no other means by which to judge a contractor's performance than visually to inspect the growing plant. Concrete's strength may be measured by an objective compression test but a plant's character has no comparable test of performance.

Unfortunately the use of common procedural planting specifications may violate basic doctrines of construction law and fairness. If a conflict between a contractor and an owner's agent arises, it may be difficult to defend a contract that directs a contractor's procedures while simultaneously holding the contractor responsible for the successful growth of plants.

Legal and fairness doctrines ordinarily view procedural specifications as obligating a contractor to only that which is described and directed by contract documents. A contractor is not considered bound simultaneously by directives of methods and procedures and for warranting or guaranteeing the success of a product. Legal and fairness doctrines generally hold that an owner who requires a contractor to warrant the success of a product must also allow a contractor to determine the methods and procedures necessary to achieve a successful product.

It may be argued that a planting contractor enters into an agreement with eyes wide open and is functioning under a format commonly and traditionally used within a segment of the construction industry. Whether or not this argument survives may well depend upon the character and language of the drawings and technical specifications as a whole. A counter argument might be that if specified procedures for installation were followed, plants were selected and located by the designer and not the contractor, reasonable care was given after planting, and the plants died, a contractor should not be held liable for a designer's decision. Such an argument would imply that the contract was impossible to execute, and thus invalid.

A key issue to be resolved in the language of a traditional procedural planting specification will be whether or not the language allows a contractor sufficient protection and involvement in planting procedures. Sufficient involvement might be defined as that which allows a contractor fairly and in good faith to accept responsibility for the guaranteeing of plantings. In other words, does a contractor have sufficient latitude to disclaim some responsibility, negotiate conflicts, verify, reduce risk, or otherwise secure a meeting of the minds and enter into an agreement with eyes wide open?

Today's planting contracts function under the present procedural specifications and guarantee format without sufficient legal conflict to change the procedure. However, a designer might assess the degree to which a contractor must function with a contingency to cover what might be construed as a transference of professional responsibility. If a contractor must view the species selected, their locations, and planting procedures as a high-risk factor in light of the guarantee, an owner may be really paying for a designer's incompetence.

7.12   THE PLANT SCHEDULE

Traditionally a plant schedule accompanies, and is intended to complement, the planting plan. There is a chance for contractual conflict between the schedule and drawings. (See Figure 7.2.)

If a plant schedule includes the quantity of each species, such information may, inadvertently conflict with a count taken from a drawing. When a conflict exists, it is not always clear as to what the contractor's obligation is. In accepted tradition, the drawing is considered to take precedence over a schedule's information. However, it is preferred that the schedule carry a provision that clarifies, for example, the fact that "... the plant schedule is given for the convenience of the contractor and is not a part of the contract documents. Quantities of plants shall be taken from the drawing information ..." This is particularly true of lump sum contracts intended to make a contractor responsible for plant quantities.

Unit price agreements automatically contain unit prices along with their respective quantity of units. Variation in such quantities is an accepted procedure and need not be clarified by a provision.

A traditional plant schedule contains the botanical plant name, common name, size, and remarks relative to special characteristics and planting instructions. The schedule might also contain maintenance practices, plant sources, guarantee information, or other information pertaining to a particular plant species.

7.13   CONSTRUCTION VERSUS HORTICULTURAL ENVIRONMENT

The relationship between general construction activities and planting is not always harmonious and can prove disastrous for plants. Problems usually arise whenever various speciality contractors are working near or in an area to be planted. Most conflict will involve construction activities that are executed without thinking about the plant environment. In many ways, contractors operate in units of time and, unless placed on notice, are not aware of the problems created by their work. For example, there is the plastering contractor who buries waste plaster and rinse water and is gone when a planting contractor discovers the debris, or the paving contractor who seal-coats asphalt on a hot, windy day and kills all plants located downwind of the work site.

Many of these conditions are preventable by simply scheduling work before or after planting operations. Contractually each contractor should be placed on notice regarding responsibilities and financial obligations for plant materials. The following list is indicative of the number of activities occurring on a typical site and the possible conditions that may affect plant health.

Possible Sources of Problems, by CSI Subdivision

01562   *Dust control.*   Control use of calcium chloride and oil.

01564   *Pest control.*   Avoid use of chemicals harmful to plants.

01620   *Storage and protection.*   Avoid soil compaction in areas to be planted as well as near existing plants to remain.

01710   *Cleaning.*   Avoid contamination of soils or plants by cleaning chemicals.

02110   *Clearing.*   Identify cleared and undisturbed vegetation.

02112   *Tree pruning.*   Avoid pruning by untrained personnel.

02481   *Shrub and tree relocation.*   Avoid work by untrained personnel (tie to guarantee of plant life).

02210   *Site grading.*   Confine equipment oiling and maintenance to one place, with soil free of contaminants or removal after grading is completed:

Avoid compaction in planted areas;
Protect existing vegetation from equipment damage;
Provide for penalty for damage to vegetation;
Avoid or repair torn roots and branches;
Arrange for maintenance of soil moisture conditions disturbed by grading or changes in drainage patterns.

02110   *Waste material disposal.*   Enforce provisions for off-site disposal of all debris and waste construction materials during and prior to planting operations; avoid burning too near existing trees and shrubs.

02240   *Soil stabilization.*   Confine specifications to unplanted areas or coordinate with planting specifications.

02281   *Termite control.*   Avoid oil-borne treatment of soil to be planted and contamination of plant foliage.

02500   *Vegetation control.*   Confine to unplanted areas, coordinate work with planting specifications, and avoid permanent soil sterilants in planted areas.

02411   *Foundation drainage.*   Avoid saturated soils or erosion at drainage discharge point.

02401   *Dewatering.*   Avoid soil saturation at

discharge points and an accumulation of toxic waters or destruction of soil texture.

02270   ***Erosion control.***   Coordinate specifications and work with planting specifications. Coordinate temporary control measures with planting protection during guarantee period. Coordinate silt traps or temporary measures with existing conditions of planting contract and finish grading.

02513   ***AC paving.***   Avoid contamination of soil from storm runoff and oil residue. Avoid overspray of oils or solvents into planted or areas to be planted. Avoid dumping of cleanings or materials into planted areas.

02515   ***PC concrete paving.***   Avoid cleaning of tools, mix trucks, mixers, acid washes, extraneous cement, and other residue into planted areas. Remove contaminated soils off site.

02576   ***Pavement sealing.***   Avoid all contamination of soils by sealant chemicals and solvents.

02463   ***Railroad tie structures.***   Avoid long-term storage or creosote-saturated railroad ties in areas to be planted.

02484   ***Soil preparation.***   Double check the application rates for all fertilizers to be applied to prevent potential nitrogen contamination.

04110   ***Cement and lime mortars.***   Remove soil contaminated with concentrations of spilled lime (depending on regional soils).

04200   ***Unit masonary.***   Avoid contamination of soils by treatments of efflorescence.

06300   ***Wood treatment.***   Avoid dripping or leaching of preservatives into planted areas, usually associated with paint-on applications.

07100   ***Waterproofing.***   Avoid the burying of membranes or other materials affecting soil drainage.

09910   ***Exterior painting.***   Avoid contamination of soils by solvents. Avoid spattering of existing plants with paint.

7.14   PROTECTION OF EXISTING PLANT MATERIALS

Several legal issues and areas of privity must be considered in developing means of protecting existing trees or other vegetation. The two common approaches involve penalties and liquidated damages. Penalty language simply means listing an amount payable to an owner in the event that a contractor damages, disfigures, or kills a particular plant. Usually a sliding scale is developed with a value based upon, for example, caliper-measured 6 inches above grade or diameter at breast height (DBH). If liquidated damages are presumed, the language will be something like this: "The actual damage to the owner will be impossible to determine and in lieu therefore, the contractor shall pay a sum to the owner as liquidated damages for each plant judged to be injured."

Under a penalty format, a value should be factually determinable and reflect *actual* costs to replace a *specific* size plant in a *specific* place. Only the use of liquidated damage language might include actual costs as well as reasonable but unquantifiable costs, such as esthetic qualities, that may be attached to a specific plant in a specific location.

An issue of privity or contractual obligations will occasionally arise in the protection of existing site features. For example, it does little good to write protection language into a prime contract for planting when a separate prime contractor may do all the damage. Such issues must run through the contract documents of *all* prime designers and prime contractors.

7.15   TOPSOIL

Whenever a designer is faced with the problem of writing a specification for topsoil, it seems easier to critically analyze an example than to write a perfect sample. The following paragraph is an example of one person's attempt to define suitable planting soil. Unfortunately the paragraph can be criticized from several angles.

Topsoil: *Topsoil shall be* natural, fertile, friable *soil possessing characteristics of representative productive soils in the vicinity. It shall be obtained from naturally* well-drained *areas. It shall not be* excessively *acid or alkaline or contain* toxic *substances which may be harmful to plant growth. Topsoil shall be without admixture of subsoil and shall be free of lumps, stones, weeds, roots, stumps, or other deleterious matter. Topsoil shall not be collected from sites that are infected with a growth of, or the reproductive parts of, noxious weeds. Topsoil shall not be stripped, collected, or deposited while wet. Emphasis added*

The words in italics, are simply undefinable in any objective or measurable form. There may be a general knowledge as to what each word means when applied to horticultural conditions, but little understanding in a *contractual* sense. Such language is of little value to a contractor in meeting contractual obligations nor is it useful to a designer who must enforce contractual intent.

Such terms as *lumps, stones, weeds, roots, stumps, deleterious, excessively, well drained, natural, fertile, subsoil, noxious,* and *wet* must be supported by means of defining contractual obligations. For instance, *weeds* and *noxious weeds* have two different meanings in common and legal use. We must assume a weed to be any plant that a contractor did not plant. However, many weeds are brought to a site via purchased seeds and, unless the contract requires temporary sterilization of the soil, are present in any soil. Is the contractor obligated to obtain a weedfree soil or later remove weeds that sprout? How would a weedfree requirement affect the establishment and maintenance periods? Such a requirement for a weedfree ordinary soil is unenforceable by contract or horticultural fact. A noxious weed is a legal term given substance by state laws that list primary and secondary noxious weeds. It is possible that a noxious weed not only may be a problem, but against the law to transport, import, or maintain.

The majority of the terms, as used in the example paragraph, suffer from an implied perfection that will produce an impossible contract under ordinary site work, where it is impossible to import or obtain a soil that is perfectly free of lumps, stones, weeds, roots, deleterious matter, and subsoil, and that is natural fertile or not wet. It is the specification writer's obligation to state, in at least explanatory terms, a range of size, quantities, or tests necessary to define terminology.

Only the first sentence of the sample paragraph is a positive statement. Unfortunately each of the succeeding sentences is negative, that is, they explain what an acceptable topsoil shall not be. In a very broad sense, negative lists are generally viewed as being complete and everything not listed then becomes acceptable. Negative language should be avoided in all forms throughout all technical specifications.

7.15.1 Imported Topsoil

Assuming that a contract requires the importation of soil as site fill, planter fill, and the like, specifications ordinarily need to recognize locally available sources, in situ conditioning, or special soil preparation.

Locally Available Soils

Technical specifications may control a soil's physical properties through reference to USDA (United States Department of Agriculture) criteria. Each soil is classified by name and content of sand, silt, and clay particles as a percent of the whole. The use of such a system controls only a soil's physical properties and not its fertility, organic matter, or moisture-holding capacity. For example, Table 2 indicates percent of particle ranges and, in parentheses, the ordinary mid percentage for representative soil types.

In Situ Conditioning

Existing or imported soils are conditioned in place by the addition of organic materials, fertilizers, minerals, sand, silts, clay, or general cultivation. A specification must include the following.

1. *Manufactured* materials must include the material, grade, size, particle size, quality, quantity, possibly the state requirements, and similar characteristics.
2. Means and methods of *incorporating* materials into a soil profile are customarily normal to landscape work. However, as previously discussed, such contractual language may cloud a guarantee of plant life. The contractor may be given several alternative methods of incorporation or otherwise become involved with demonstrating methodology to fit written performance criteria.

Specifications must set criteria as to the quantity of material to be incorporated per

TABLE 2

	Percentage Sand[a]	Percentage Silt[b]	Percentage Clay[c]
Sand	85−100(90)	0−10(5)	0−10(5)
Loamy sand	70−85(80)	0−30(15)	0−15(5)
Sandy loam	50−80(65)	0−50(25)	0−20(10)
Loam	25−52(40)	28−50(40)	8−28(20)
Silt loam	0−50(20)	50−88(65)	0−28(15)
Sandy silt loam	45−80(60)	0−28(15)	20−35(25)
Clay loam	20−45(35)	15−52(30)	28−40(35)
Silty clay loam	0−20(10)	40−70(55)	28−40(35)
Sandy clay	45−63(55)	0−20(5)	35−57(45)
Silty clay	0−20(5)	40−60(45)	40−60(55)
Clay	0−45(20)	0−40(20)	40−100(60)

[a]Sand: particles 2.0−0.05 millimeter in diameter
[b]Silt: particles 0.05−0.002 millimeter in diameter
[c]Clay: particles below 0.002 millimeter in diameter

unit of area, the depth of incorporation from finish grade, and a reasonable distribution pattern of the material in the soil particles. The relationship of incorporation timing and methods to dust, soil moisture condition, finish grade completion, completion of excavations, noise from equipment, protection of below-grade root zones and obstacles, sprinkler heads, and the like, must be considered.

Special Preparation

On occasion, a designer may wish to specify specific areas, containers, or plant backfill as requiring special preparation. In some instances, this concern may involve sands to maximize drainage through a medium's profile, a mixture of topsoil and various admixtures, a lightweight growing medium for rooftop or other dead-load or a live-load conditions, or sterlization.

Each special mixture must be specified as to the proportions of its constituents and either method or performance criteria for mixing. In most cases, a designer must decide on the degree of uniformity to be sought in the final mixture. For example, if specialized constituents are mixed in large rotating drums, as concrete might be, it is possible to obtain a very uniform distribution of particles. On the other hand, onsite mixing with rototilling, windrow and road construction equipment, and similar specialized or adapted equipment may be suitable

and less expensive, and may provide acceptable particle distribution. In all instances, the drawings and details must be very explicit as to the area or installation to receive a special mixture, depth, compaction, and finish grade in order to avoid confusion with ordinary soil.

Contractual Options

Existing soil may be tested as a part of the owner's responsibility for existing site information (see Chapter 2). From such horticultural information, a designer may formulate a contractor's scope of lump sum or unit cost work for importation, in situ conditioning, or special preparation. The type of admixtures, quantities required, areas covered, and methodology of incorporation are developed from the known data of the existing or borrow soil.

If it is not possible to ascertain the nature or location of proposed borrow soil, it will not be possible to write knowledgeable specifications directing contractual obligations. Assuming that a technical specification controls only the physical nature of a borrow soil, a test of such soil can be made after a contractor has been selected. Several directions can be taken by the specification language. First, the contractual specifications can be made quite explicit as to the physical, organic, and nutrient value of a borrow soil. Second, a test may be made on the borrow to check its acceptability and conformance. Third, the contractual specifications can control only the physical nature of the borrow soil. Fourth, the borrow may be tested as a check for physical acceptability and compliance and to gather data as to the organic and nutrient content of the soil. Fifth, the contractor may then be directed to condition the borrow by modifying the contractor−owner agreement through a change order, or acceptance of an alternate or unit price in the bid or negotiation.

SELECTED READING AND REFERENCES

American Association of Nurserymen, *American Standard for Nursery Stock*. Document ANSI-Z60.1—1973. Washington, D.C.: American Association of Nurserymen.

National Arborist Association *Standards for Pruning, Guying, Fertilizing, Lighting Protection and Spraying Operation for Shade and Ornamental Trees*, Wantagh, New York: National Arborist Association, 1979.

Robinette, Gary O. *Off the Board/Into the Ground: Techniques of Planting Design Implementation.* Dubuque, Iowa: W. C. Brown, 1968.

Conover, Herbert S. *Grounds Maintenance Handbook.* New York: McGraw-Hill, 1977.

Hannebau, Leroy G. *Landscape Operations, Management, Methods and Materials.* Reston, Virginia: Reston, 1980.

Carpenter, Philip L., et al. *Plants in the Landscape.* San Francisco: W. H. Freeman, 1975.

American Joint Committee on Horticultural Nomenclature *Standardized Plant Names,* 2nd ed. Harrisburg, Pennsylvania: J. Horace McFarland, 1942.

Index